REA's Books Are

They have rescued lots of

(a sample of the <u>hundreds of letters</u> REA receives each year)

"Your books are great! They are very helpful, and have upped my grade in every class. Thank you for such a great product."

Student, Seattle, WA

"Your book has really helped me sharpen my skills and improve my weak areas. Definitely will buy more."

Student, Buffalo, NY

"Compared to the other books that my fellow students had, your book was the most useful in helping me get a great score."

Student, North Hollywood, CA

"I really appreciate the help from your excellent book. Please keep up your great work."

Student, Albuquerque, NM

"Your book was such a better value and was so much more complete than anything your competition has produced (and I have them all)!"

Teacher, Virginia Beach, VA

(more on next page)

(continued from previous page)

"Your books have saved my GPA, and quite possibly my sanity.
My course grade is now an 'A', and I couldn't be happier."

Student, Winchester, IN

"These books are the best review books on the market.
They are fantastic!"

Student, New Orleans, LA

"Your book was responsible for my success on the exam. . . I
will look for REA the next time I need help."

Student, Chesterfield, MO

"I think it is the greatest study guide I have ever used!"

Student, Anchorage, AK

"I encourage others to buy REA because of their superiority.
Please continue to produce the best quality books on the market."

Student, San Jose, CA

"Just a short note to say thanks for the great support your book
gave me in helping me pass the test . . . I'm on my way to a
B.S. degree because of you !"

Student, Orlando, FL

Super Review™

All You Need to Know!

GEOMETRY

By the Staff of
Research & Education Association
Dr. M. Fogiel, Director

Research & Education Association
61 Ethel Road West
Piscataway, New Jersey 08854

SUPER REVIEW ™
OF GEOMETRY

Printed in the United States of America

Library of Congress Catalog Card Number 00-130290

International Standard Book Number 0-87891-188-X

SUPER REVIEW is a trademark of
Research & Education Association, Piscataway, New Jersey 08854

WHAT THIS Super Review WILL DO FOR YOU

This **Super Review** provides all that you need to know to do your homework effectively and succeed on exams and quizzes.

The book focuses on the core aspects of the subject, and helps you to grasp the important elements quickly and easily.

Outstanding **Super Review** features:

- Topics are covered in logical sequence

- Topics are reviewed in a concise and comprehensive manner

- The material is presented in student-friendly language that makes it easy to follow and understand

- Individual topics can be easily located

- Provides excellent preparation for midterms, finals and in-between quizzes

- In every chapter, reviews of individual topics are accompanied by Questions **Q** and Answers **A** that show how to work out specific problems

- At the end of most chapters, quizzes with answers are included to enable you to practice and test yourself to pinpoint your strengths and weaknesses

- Written by professionals and test experts who function as your very own tutors

Dr. Max Fogiel
Program Director

CONTENTS

8   GEOMETRIC PROPORTIONS
   AND SIMILARITY

CHAPTER 1

Method of Proof

1.1 Logic

Definition 1

A statement is a sentence which is either true or false, but not both.

Definition 2

If a and b are statements, then a statement of the form "a and b" is called the conjunction of a and b, denoted by $a \wedge b$.

Definition 3

The disjunction of two statements a and b is shown by the compound statement "a or b," denoted by $a \vee b$.

Definition 4

The negation of a statement q is the statement "not q," denoted by $\sim q$.

Definition 5

The compound statement "if a, then b," denoted by $a \to b$, is called a conditional statement or an implication.

"If a" is called the hypothesis or premise of the implication, "then b" is called the conclusion of the implication.

Further, statement a is called the antecedent of the implication, and statement b is called the consequent of the implication.

Definition 6

The converse of $a \to b$ is $b \to a$.

Definition 7

The contrapositive of $a \to b$ is $\sim b \to \sim a$.

Definition 8

The inverse of $a \to b$ is $\sim a \to \sim b$.

Definition 9

The statement of the form "p if and only if q," denoted by $p \leftrightarrow q$, is called a biconditional statement.

Definition 10

An argument is valid if the truth of the premises means that the conclusion must also be true.

Definition 11

Intuition is the process of making generalizations on insight.

Problem Solving Examples:

Q Write the inverse for each of the following statements. Determine whether the inverse is true or false. (a) If a person is stealing, he is breaking the law. (b) If a line is perpendicular to a segment at its midpoint, it is the perpendicular bisector of the segment. (c) Dead men tell no tales.

A The inverse of a given conditional statement is formed by negating both the hypothesis and conclusion of the conditional statement.

(a) The hypothesis of this statement is "a person is stealing"; the conclusion is "he is breaking the law." The negation of the hypothesis is "a person is not stealing." The inverse is "if a person is not stealing, he is not breaking the law."

The inverse is false, since there are more ways to break the law than by stealing. Clearly, a murderer may not be stealing but he is surely breaking the law.

(b) In this statement, the hypothesis contains two conditions: (1) the line is perpendicular to the segment; and (2) the line intersects the segment at the midpoint. The negation of (statement a *and* statement b) is (not statement a *or* not statement b). Thus, the negation of the hypothesis is "The line is not perpendicular to the segment or it doesn't intersect the segment at the midpoint." The negation of the conclusion is "the line is not the perpendicular bisector of a segment."

The inverse is "if a line is not perpendicular to the segment or does not intersect the segment at the midpoint, then the line is not the perpendicular bisector of the segment."

In this case, the inverse is true. If either of the conditions holds (the line is not perpendicular; the line does not intersect at the midpoint), then the line cannot be a perpendicular bisector.

(c) This statement is not written in if-then form, which makes its hypothesis and conclusion more difficult to see. The hypothesis is implied to be "the man is dead"; the conclusion is implied to be "the man tells no tales." The inverse is, therefore, "If a man is not dead, then he will tell tales."

The inverse is false. Many witnesses to crimes are still alive but they have never told their stories to the police, either out of fear or because they didn't want to get involved.

Q Write the converse of the following statement. Consider whether the converse is true or false. If two triangles are congruent, they have three pairs of congruent angles.

A The converse of a given statement is another statement which is formed by interchanging the hypothesis and the conclusion in the given statement. If the statement is true, the converse may be true but does not necessarily have to be.

In this statement, "two triangles are congruent" is the hypothesis and, "they have three pairs of congruent angles" is the conclusion. The converse of the original statement is, "If two triangles have three pairs of congruent angles, they are congruent."

The only methods for proving congruence between triangles are sss $\cong$ sss, a $\cdot$ s $\cdot$ a $\cong$ a $\cdot$ s $\cdot$ a, a $\cdot$ a $\cdot$ s $\cong$ a $\cdot$ a $\cdot$ s and s $\cdot$ a $\cdot$ s $\cong$ s $\cdot$ a $\cdot$ s. Having three pairs of congruent angles is a proof for triangle similarity, not congruence. Therefore, the converse is false.

Basic Principles, Laws, and Theorems

1. Any statement is either true or false. (The Law of the Excluded Middle)

2. A statement cannot be both true and false. (The Law of Contradiction)

3. The converse of a true statement is not necessarily true.

4. The converse of a definition is always true.

5. For a theorem to be true, it must be true for all cases.

6. A statement is false if one false instance of the statement exists.

7. The inverse of a true statement is not necessarily true.

8. The contrapositive of a true statement is true and the contrapositive of a false statement is false.

9. If the converse of a true statement is true, then the inverse is true. Likewise, if the converse is false, the inverse is false.

10. Statements which are either both true or false are said to be logically equivalent.

11. If a given statement and its converse are both true, then the conditions in the hypothesis of the statement are both necessary and sufficient for the conclusion of the statement.

  If a given statement is true but its converse is false, then the conditions are sufficient but not necessary for the conclusion of the statement.

  If a given statement and its converse are both false, then the conditions are neither sufficient nor necessary for the statement's conclusion.

1.2 Deductive Reasoning

An arrangement of statements that would allow you to deduce the third one from the preceding two is called a syllogism. A syllogism has three parts:

The first part is a general statement concerning a whole group. This is called the major premise.

The second part is a specific statement which indicates that a certain individual is a member of that group. This is called the minor premise.

The last part of a syllogism is a statement to the effect that the general statement which applies to the group also applies to the individual. This third statement of a syllogism is called a deduction.

Example A: Properly Deduced Argument

A) Major Premise: All birds have feathers.

B) Minor Premise: An eagle is a bird.

C) Deduction: An eagle has feathers.

The technique of employing a syllogism to arrive at a conclusion is called deductive reasoning.

If a major premise which is true is followed by an appropriate minor premise which is true, a conclusion can be deduced which must be true, and the reasoning is valid. However, if a major premise which is true is followed by an inappropriate minor premise which is also true, a conclusion cannot be deduced.

Example B: Improperly Deduced Argument

A) Major Premise: All people who vote are at least 18 years old.

B) Improper Minor Premise: Jane is at least 18.

C) Illogical Deduction: Jane votes.

The flaw in example B is that the major premise stated in A makes a condition on people who vote, not on a person's age. If statements B and C are interchanged, the resulting three-part deduction would be logical.

1.3 Indirect Proof

Indirect proofs involve considering two possible outcomes—the result we would like to prove and its negative—and then showing, under the given hypothesis, that a contradiction of prior known theorems, postulates, or definitions is reached when the negative is assumed.

Postulate 1

A proposition contradicting a true proposition is false.

Postulate 2

If one of a given set of propositions must be true, and all except one of those propositions have been proved to be false, then this one remaining proposition must be true.

The method of indirect proof may be summarized as follows:

Step 1. List all the possible conclusions.

Step 2. Prove all but one of those possible conclusions to be false (use Postulate 1 given).

Step 3. The only remaining possible conclusion is proved true according to Postulate 2.

Example

When attempting to prove that in a scalene triangle the bisector of an angle cannot be perpendicular to the opposite side, one method of solution could be to consider the two possible conclusions:

1) the bisector can be perpendicular to the opposite side, or

2) the bisector cannot be perpendicular to the opposite side.

Obviously, one and only one of these conclusions can be true; therefore, if we can prove that all of the possibilities, except one, are false, then the remaining possibility must be a valid conclusion. In this example, it can be proven that, for all cases, the statement which asserts that the bisector of an angle of a scalene triangle can be perpendicular to the opposite side is false. Therefore, the contradicting possibility—the bisector cannot be perpendicular to the opposite side—is in fact true.

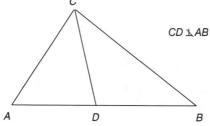

$CD \perp AB$

Problem Solving Example:

 Prove, by indirect method, that if two angles are not congruent, then they are not both right angles.

 Indirect proofs involve considering two possible outcomes, the result we would like to prove and its negative, and then showing, under the given hypothesis, that a contradiction of prior known theorems, postulates, or definitions is reached when the negative is assumed.

In this case, the outcomes can be that the two angles are not right angles or that the two angles are right angles. Assume the negative of what we want to prove—that the two angles are right angles.

The given hypothesis in this problem is that the two angles are not congruent. A previous theorem states that all right angles are congruent. Therefore, the conclusion we have assumed true leads to a logical contradiction. As such, the alternative conclusion must be true. Therefore, if two angles are not congruent, then they are not both right angles.

1.4   Inductive Reasoning

Definition 1

Inductive reasoning is a method of reasoning in which one draws conclusions or generalizations from several known particular cases. The resulting conclusion is called an induction.

Definition 2

Inductive reasoning is the drawing of a conclusion based on experimenting with particular examples. A typical mathematical induction is the Method of Proof whereby:

   A)   the conditions of the statement are valid for the smallest possible value of n;

B) the conditions are assumed for the general case, say $n = k$;

C) the conditions are tested and verified for $n = k + 1$.

If A is proved true, and C is true when B is assumed, then the statement is valid for all n greater than or equal to the smallest n.

Problem Solving Example:

 Prove by mathematical induction that
$$1 + 7 + 13 + \ldots + (6n - 5) = n(3n - 2).$$

 (1) The proposed formula is true for $n = 1$, since $1 = 1(3 - 2)$.
(2) Assume the formula to be true for $n = k$, a positive integer; that is, assume

(A) $1 + 7 + 13 + \ldots + (6k - 5) = k(3k - 2)$.

Under this assumption, we wish to show that

(B) $1 + 7 + 13 + \ldots + (6k - 5) + (6k + 1) = (k + 1)(3k + 1)$.

When $(6k + 1)$ is added to both members of (A), we have on the right

$$k(3k - 2) + (6k + 1) = 3k^2 + 4k + 1 = (k + 1)(3k + 1);$$

hence, if the formula is true for $n = k$, it is true for $n = k + 1$.

(3) Since the formula is true for $n = k = 1$ (Step 1), it is true for $n = k + 1 = 2$; being true for $n = k = 2$, it is true for $n = k + 1 = 3$; and so on, for every positive integral value of n.

1.5 Defined and Undefined Terms: Axioms, Postulates, and Assumptions; Theorems and Corollaries

To build a logical system of mathematics, the first step is to take a known and then move to what is not known. The terms which we will accept as known are called undefined terms. We accept certain basic terms as undefined, since their definition would of necessity include other undefined terms. Examples of some important undefined terms with characteristics that you must know are:

A) Set: The sets we will be concerned with will have clearly defined characteristics.

B) Point: Although we represent points on paper with small dots, a point has no size, thickness, or width. A point is denoted by a capital letter.

C) Line: A line is a series of adjacent points which extends indefinitely. A line can be either curved or straight; however, unless otherwise stated, the term "line" refers to a straight line. A line is denoted by listing two points on the line and drawing a line with arrows on top, i.e., $\overleftrightarrow{AB}$.

D) Plane: A plane is the collection of all points lying on a flat surface which extends indefinitely in all directions. Imagine holding a record cover in a room and imagine that the record cover divides the entire room. Remember that a plane has no thickness.

We use these undefined terms to construct defined terms so we can describe more sophisticated expressions.

Necessary characteristics of a good definition are:

A) It names the term being defined.

B) It uses only known terms or accepted undefined terms.

C) It places the term into the smallest set to which it belongs.

D) It states the characteristics of the defined term which distinguish it from the other members of the set.

E) It contains the least possible amount of information.

F) It is always reversible.

Axioms, postulates, and assumptions are the statements in geometry which are accepted as true without proof, whereas theorems are the statements in geometry which are proven to be true.

A corollary is a theorem that can be deduced easily from another theorem or from a postulate.

In this text, the term postulate is used exclusively, instead of axiom or assumption.

Postulate 1

A quantity is equal to itself (reflexive law).

Postulate 2

If two quantities are equal to the same quantity, they are equal to each other (transitive law).

Postulate 3

If a & b are any quantities, and $a = b$, then $b = a$ (symmetric law).

Postulate 4

The whole is equal to the sum of its parts.

Postulate 5

If equal quantities are added to equal quantities, the sums are equal quantities.

Postulate 6

If equal quantities are subtracted from equal quantities, the differences are equal quantities.

Postulate 7

If equal quantities are multiplied by equal quantities, the products are equal quantities.

Postulate 8

If equal quantities are divided by equal quantities (not 0), the quotients are equal quantities.

Postulate 9

There exists one and only one straight line through any two distinct points.

Postulate 10

Two straight lines can intersect at only one point.

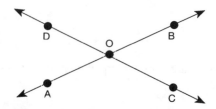

Problem Solving Examples:

Q In the figure shown, the measure of ∢*DAC* equals the measure of ∢*ECA* and the measure of ∢1 equals the measure of ∢2. Show that the measure of ∢3 equals the measure of ∢4.

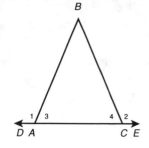

This proof will require the subtraction postulate, which states that if equal quantities are subtracted from equal quantities, the differences are equal.

Given: $\angle DAC \cong \angle ECA$, $\angle 1 \cong \angle 2$

Prove: $\angle 3 \cong \angle 4$

Statement	Reason
1. $m \angle DAC = m \angle ECA$ $m \angle 1 = m \angle 2$	1. Given.
2. $m \angle DAC - m \angle 1 = m \angle ECA - m \angle 2$	2. Subtraction Postulate.
3. $m \angle 3 = m \angle 4$	3. Substitution Postulate.

 In the diagram $AB = CD$, $RS = 2AB$, and $LM = 2CD$. Prove that $RS = LM$.

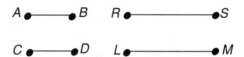

$A \bullet\!\!-\!\!-\!\!-\!\!\bullet B \qquad R \bullet\!\!-\!\!-\!\!-\!\!-\!\!-\!\!\bullet S$

$C \bullet\!\!-\!\!-\!\!\bullet D \qquad L \bullet\!\!-\!\!-\!\!-\!\!-\!\!\bullet M$

 This proof will involve an application of the Multiplication Postulate, which states that if equal quantities are multiplied by equal quantities, the products are equal.

Statement	Reason
1. $AB = CD$	1. Given.
2. $2AB = 2CD$	2. Multiplication Postulate.
3. $RS = 2 \cdot AB$ $LM = 2 \cdot CD$	3. Given.
4. $RS = LM$	4. Doubles of equal quantities are equal.

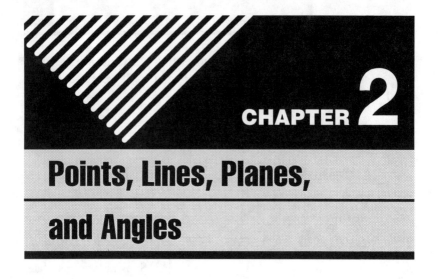

CHAPTER 2

Points, Lines, Planes, and Angles

2.1 Definitions

Definition 1

If A and B are two points on a line, then the line segment AB is the set of points on that line between A and B and including A and B, which are called the endpoints. The line segment is referred to as $\overline{AB}$.

$A \qquad B$

Definition 2

A half-line is the set of all the points on a line on the same side of a dividing point, not including the dividing point, denoted by $\overrightarrow{AB}$.

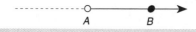

$A \qquad B$

Definition 3

Let A be a dividing point on a line. Then, a ray is the set of all the points on a half-line and the dividing point itself. The dividing point is called the endpoint or the vertex of the ray. The ray AB shown be-low is denoted by $\overrightarrow{AB}$.

Definition 4

Three or more points are said to be collinear if and only if they lie on the same line.

Definition 5

Let X, Y, and Z be three collinear points. If Y is between X and Z, then $\overrightarrow{YX}$ and $\overrightarrow{YZ}$ are called opposite rays.

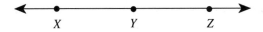

Definition 6

The absolute value of x, denoted by $|x|$ is defined as

$$|x| = \begin{cases} x & \text{if} \quad x > 0 \\ 0 & \text{if} \quad x = 0 \\ -x & \text{if} \quad x < 0 \end{cases}$$

Definition 7

The absolute value of the difference of the coordinates of any two points on the real number line is the distance between those two points.

Definition 8

The length of a line segment is the distance between its endpoints.

Definition 9

Congruent segments are segments that have the same length.

Definition 10

The midpoint of a segment is defined as the point of the segment which divides the segment into two congruent segments. (The midpoint is said to bisect the segment.)

Problem Solving Examples:

Solve for x when $|x - 7| = 3$.

This equation, according to the definition of absolute value, expresses the conditions that $x - 7$ must be 3 or -3, since in either case the absolute value is 3. If $x - 7 = 3$, we have $x = 10$; and if $x - 7 = -3$, we have $x = 4$. We see that there are two values of x which solve the equation.

Find point C between A and B in the figure below such that $\overline{AC} \cong \overline{CB}$.

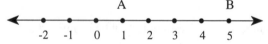

We must determine point C in such a way that $\overline{AC} \cong \overline{CB}$, or $AC = CB$. We are first given that C is between A and B. Therefore, since the measure of the whole is equal to the sum of the measure of its parts:

(I) $AC + CB = AB$

Using these two facts, we can find the length of AC. From that we can find C.

First, since $AC = CB$, we substitute AC for CB in equation (I)

(II) $AC + AC = AB$

(III) $2(AC) = AB$

Dividing by 2 we have

(IV) AC = 1/2 AB

To find AC, we must know AB. We can find AB from the coordinates of A and B. They are 1 and 5, respectively. Accordingly,

(V) AB = $|5 - 1|$

(VI) AB = 4

We substitute 4 for AB in equation (IV)

(VII) AC = 1/2 (4)

(VIII) AC = 2.

Therefore, C is 2 units from A. Since C is between A and B, the coordinate of C must be 3.

Definition 11

The bisector of a line segment is a line that divides the line segment into two congruent segments.

Definition 12

An angle is a collection of points which is the union of two rays having the same endpoint. An angle such as the one illustrated in Figure 1 can be referred to in any of the following ways:

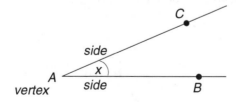

A) by a capital letter which names its vertex, i.e., ∢ A;

B) by a lowercase letter or number placed inside the angle, i.e., ∢x;

C) by three capital letters, where the middle letter is the vertex and the other two letters are not on the same ray, i.e., ∡*CAB* or ∡*BAC*, both of which represent the angle illustrated in Figure 1.

Definition 13

A set of points is coplanar if all the points lie in the same plane.

Definition 14

Two angles with a common vertex and a common side, but no common interior points, are called adjacent angles.

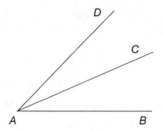

In the above figure, ∡ DAC and ∡ BAC are adjacent angles; ∡ DAB and ∡ BAC are not adjacent angles.

Definition 15

Vertical angles are two angles with a common vertex and with sides that are two pairs of opposite rays.

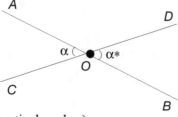

(∡ α and ∡ α* are vertical angles.)

Definition 16

An acute angle is an angle whose measure is larger than 0° but smaller than 90°.

Definition 17

An angle whose measure is 90° is called a right angle.

Definition 18

An obtuse angle is an angle whose measure is larger than 90° but less than 180°.

Definition 19

An angle whose measure is 180° is called a straight angle. Note: Such an angle is, in fact, a straight line.

Definition 20

An angle whose measure is greater than 180° but less than 360° is called a reflex angle.

Definition 21

Complementary angles are two angles, the sum of the measures of which equals 90°.

Definition 22

Supplementary angles are two angles, the sum of the measures of which equals 180°.

Definition 23

Congruent angles are angles of equal measure.

Definition 24

A ray bisects (is the bisector of) an angle if the ray divides the angle into two angles that have equal measure.

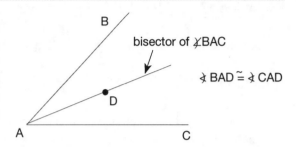

Definition 25

If the two non-common sides of adjacent angles form opposite rays, then the angles are called a linear pair. Note that α and β are supplementary.

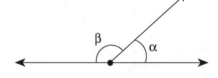

Definition 26

Two lines are said to be perpendicular if they intersect and form right angles. The symbol for perpendicular (or, is perpendicular to) is $\perp$; $\overleftrightarrow{AB}$ is perpendicular to $\overleftrightarrow{CD}$ is written $\overleftrightarrow{AB} \perp \overleftrightarrow{CD}$.

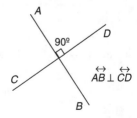

Definition 27

A line, a ray, or a line segment which bisects a line segment and is also perpendicular to that segment is called a perpendicular bisector of the line segment.

Definition 28

The distance from a point to a line is the measure of the perpendicular line segment from the point to that line. Note: This is the shortest possible distance of the point to the line, which is later stated in Chapter 7, Theorem 1.

Definition 29

Two or more distinct lines are said to be parallel (‖) if and only if they are coplanar and they do not intersect.

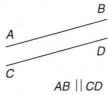

AB ‖ CD

Definition 30

The projection of a given point on a given line is the foot of the perpendicular drawn from the given point to the given line.

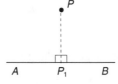

P_1 is the projection of *P* on $\overleftrightarrow{AB}$

The foot of a perpendicular from a point to a line is the point where the perpendicular meets the line.

Definition 31

The projection of a segment on a given line (when the segment is not perpendicular to the line) is a segment with endpoints that are the projections of the endpoints of the given line segment onto the given line.

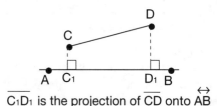

C_1D_1 is the projection of $\overline{CD}$ onto $\overleftrightarrow{AB}$

Problem Solving Examples:

 The measure of the complement of a given angle is four times the measure of the angle. Find the measure of the given angle.

 By the definition of complementary angles, the sum of the measures of the two complements must equal 90°.

Accordingly,

(1)   Let $x =$ the measure of the angle

(2)   Then $4x =$ the measure of the complement of this angle.

Therefore, from the discussion above,

$$x + 4x = 90°$$
$$5x = 90°$$
$$x = 18°$$

Therefore, the measure of the given angle is 18°.

In the figure, we are given $\overleftrightarrow{AB}$ and triangle ABC. We are told that the measure of ∡1 is five times the measure of ∡2. Determine the measures of ∡1 and ∡2.

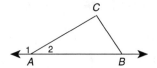

A Since ∢1 and ∢2 are adjacent angles whose non-common sides lie on a straight line, they are, by definition, supplementary. As supplements, their measures must sum to 180°.

If we let x = the measure of ∢2

then, 5x = the measure of ∢1.

To determine the respective angle measures, set $x + 5x = 180$ and solve for x. $6x = 180$. Therefore, $x = 30$ and $5x = 150$.

Therefore, the measure of ∢1 = 150 and the measure of ∢2 = 30.

2.2 Postulates

Postulate 1 (The Point Uniqueness Postulate)

Let n be any positive number, then there exists exactly one point N of $\overrightarrow{AB}$ such that $AN = n$. (AN is the length of $\overline{AN}$)

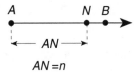

$$AN = n$$

Postulate 2 (The Line Postulate)

Any two distinct points determine one and only one line which contains both points.

Postulate 3 (The Point Betweenness Postulate)

Let A and B be any two points. Then, there exists at least one point (and in fact an infinite number of such points) of $\overleftrightarrow{AB}$ such that P is between A and B, with $AP + PB = AB$.

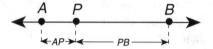

Postulate 4

Two distinct straight lines can intersect at most at only one point.

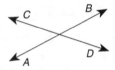

Postulate 5

The shortest line between any two points is a straight line.

Postulate 6

There is a one-to-one correspondence between the real numbers and the points of a line. That is, to every real number there corresponds exactly one point of the line and to every point of the line there corresponds exactly one real number. (In other words, a line has an infinite number of points between any two distinct points.)

Postulate 7

One and only one perpendicular can be drawn to a given line through any point on that line. Given point O on line $\overleftrightarrow{AB}$, $\overleftrightarrow{OC}$ represents the only perpendicular to AB which passes through O.

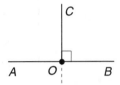

Problem Solving Examples:

Q In the accompanying figure, point B is between points A and C, and point E is between points D and F. Given that $\overline{AB} \cong \overline{DE}$ and $\overline{BC} \cong \overline{EF}$. Prove that $\overline{AC} \cong \overline{DF}$.

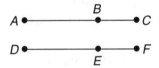

A Two important postulates will be employed in this proof. The Point Betweenness Postulate states that if point Y is between point X and Z, then $XY + YZ = XZ$. Furthermore, the Postulate states that the converse is also true—that is, if $XY + YZ = XZ$, then point Y is between point X and Z.

The Addition Postulate states that equal quantities added to equal quantities yield equal quantities. Thus, if $a = b$ and $c = d$, then $a + c = b + d$.

Given: Point B is between A and C; point E is between points D and F; $\overline{AB} \cong \overline{DE}$; $\overline{BC} \cong \overline{EF}$

Prove: $\overline{AC} \cong \overline{DF}$.

Statement	Reason
1. (For the given, see above)	1. Given.
2. $\quad AB = DE$ $\quad BC = EF$	2. Congruent segments have equal lengths.
3. $AB + BC = DE + EF$	3. Addition Postulate.
4. $\quad AC = DF$	4. Point Between Postulate.
5. $\quad \overline{AC} \cong \overline{DF}$	5. Segments of equal length are congruent.

 Construct a line perpendicular to a given line through a given point on the given line.

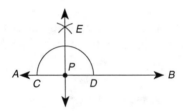

 Let line $\overset{\leftrightarrow}{AB}$ and point P be the given line and the given point, respectively.

We notice that $\angle APB$ is a straight angle. A line perpendicular to $\overset{\leftrightarrow}{AB}$ from point P will form adjacent congruent angles with $\overset{\leftrightarrow}{AB}$, by the definition of a perpendicular. Since $\angle APB$ is a straight angle, the adjacent angles will be right angles. As such, the required perpendicular is the angle bisector of $\angle APB$.

We can complete our construction by bisecting $\angle APB$.

1. Using P as the center and any convenient radius, construct an arc which intersects $\overset{\leftrightarrow}{AB}$ at points C and D.

2. With C and D as centers and with a radius greater in length than the one used in Step 1, construct arcs that intersect. The intersection point of these two arcs is point E.

3. Draw $\overleftrightarrow{EP}$.

$\overleftrightarrow{EP}$ is the required angle bisector and, as such, $\overleftrightarrow{EP} \perp \overleftrightarrow{AB}$.

Q Present a formal proof of the following conditional statement:
If $\overleftrightarrow{CE}$ bisects $\sphericalangle ADB$, and if $\overleftrightarrow{FDB}$ and $\overleftrightarrow{CDE}$ are straight lines,
then $\sphericalangle a \cong \sphericalangle x$. (Refer to the accompanying figure).

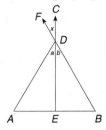

A In this problem, it will be necessary to recognize vertical angles and be knowledgeable of their key properties. Furthermore, we will need the definition of the bisector of an angle.

Vertical angles are two angles which have a common vertex, and whose sides are two pairs of opposite rays. Vertical angles are always congruent.

Lastly, the bisector of any angle divides the angle into two congruent angles.

Statement	Reason
1. $\overleftrightarrow{CE}$ bisects $\sphericalangle ADB$	1. Given.
2. $\sphericalangle a \cong \sphericalangle b$	2. A bisector of an angle divides the angle into two congruent angles.
3. $\overleftrightarrow{FDB}$ and $\overleftrightarrow{CDE}$ are straight lines	3. Given.
4. $\sphericalangle x$ and $\sphericalangle b$ are vertical angles	4. Definition of vertical angles.
5. $\sphericalangle b \cong \sphericalangle x$	5. Vertical angles are congruent.
6. $\sphericalangle a \cong \sphericalangle x$	6. Transitivity property of congruence of angles.

Note that step 3 is essential because without $\overleftrightarrow{FDB}$ and $\overleftrightarrow{CDE}$ being straight lines the definition of vertical angles would not be applicable to ⊀x and ⊀b.

Postulate 8

The perpendicular bisector of a line segment is unique.

Postulate 9 (The Plane Postulate)

Any three non-collinear points determine one and only one plane that contains those three points.

Postulate 10 (The Points-in-a-Plane Postulate)

If two distinct points of a line lie in a given plane, then the line lies in that plane.

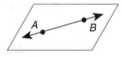

Postulate 11 (Plane Separation Postulate)

Any line in a plane separates the plane into two half planes.

Postulate 12

Given an angle, there exists one and only one real number between 0 and 180 corresponding to it. Note: m ⊀A refers to the measurement of angle A.

Postulate 13 (The Angle Sum Postulate)

If A is in the interior of $\angle XYZ$, then $m\angle XYZ = m\angle XYA + m\angle AYZ$.

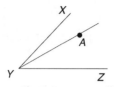

Postulate 14 (The Angle Difference Postulate)

If P is in the exterior of $\angle ABC$ and in the same half-plane (created by edge $\overleftrightarrow{BC}$) as A, then $m \angle ABP = m \angle PBC - m \angle ABC$.

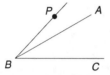

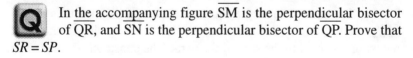

Problem Solving Examples:

Q In the accompanying figure $\overline{SM}$ is the perpendicular bisector of $\overline{QR}$, and $\overline{SN}$ is the perpendicular bisector of $\overline{QP}$. Prove that $SR = SP$.

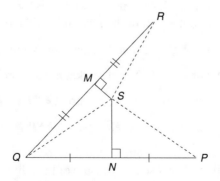

Every point on the perpendicular bisector of a segment is equidistant from the endpoints of the segment.

Since point S is on the perpendicular bisector of $\overline{QR}$,

(I) $SR = SQ$

Also, since point S is on the perpendicular bisector of $\overline{QP}$,

(II) $SQ = SP$

By the transitive property (quantities equal to the same quantity are equal), we have:

(III) $SR = SP$.

 To construct an angle whose measure is equal to the sum of the measures of two given angles.

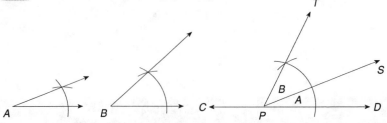

 To construct an angle equal to the sum of the measures of two given angles, we must invoke the theorem which states that the whole is equal to the sum of the parts. The construction, then, will duplicate the given angles in such a way as to form one larger angle equal in measure to the sum of the measures of the two given angles.

The two given angles, $\angle A$ and $\angle B$, are shown in the figure.

1. Construct any line $\overleftrightarrow{CD}$, and mark a point P on it.

2. At P, using $\overrightarrow{PD}$ as the base, construct $\angle DPS \cong \angle A$.

3. Now, using $\overrightarrow{PS}$ as the base, construct $\angle SPT \cong \angle B$ at point P.

4. $\angle DPT$ is the desired angle, equal in measure to $m \angle A + m \angle B$. This follows because the measure of the whole, $\angle DPT$, is equal to the sum of the measure of the parts, $\angle A$ and $\angle B$.

2.3 Theorems

Theorem 1

All right angles are equal.

Theorem 2

All straight angles are equal.

Theorem 3

Supplements of the same or equal angles are themselves equal.

Theorem 4

Complements of the same or equal angles are themselves equal.

Theorem 5

Vertical angles are equal.

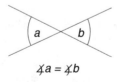

$$\angle a = \angle b$$

Theorem 6

Two supplementary angles are right angles if they have the same measure.

$$\alpha = \beta = 90°$$

Theorem 7

If two lines intersect and form one right angle, then the lines form four right angles.

Problem Solving Examples:

 Find the measure of the angle whose measure is 40° more than the measure of its supplement.

 By the definition of supplementary angles, the sum of the measures of two supplements must equal 180°. Accordingly,

Let x = the measure of the supplement of the angle.

Then $x + 40°$ = the measure of the angle.

Therefore, $$x + (x + 40°) = 180°$$
$$2x + 40° = 180°$$
$$2x = 140°$$
$$x = 70° \text{ and } x + 40° = 110°.$$

Therefore the measure of the angle is 110°.

 What is the measure of a given angle whose measure is half the measure of its complement?

 When two angles are said to be complementary we know that their measures must sum, by definition, to 90°.

If we let x = the measure of the given angle,

then $2x$ = the measure of its complement.

To determine the measure of the given angle, set the sum of the two angle measures equal to 90 and solve for x. Accordingly,

$$x + 2x = 90$$
$$3x = 90$$
$$x = 30$$

Therefore, the measure of the given angle is 30° and its complement is 60°.

Q Given that straight lines $\overleftrightarrow{AB}$ and $\overleftrightarrow{CD}$ intersect at point E, that $\angle BEC$ has measure 20° greater than 5 times a fixed quantity, and that $\angle AED$ has measure 60° greater than 3 times this same quantity: Find a) the unknown fixed quantity, b) the measure of $\angle BEC$, and c) the measure of $\angle CEA$. (For the actual angle placement, refer to the accompanying diagram.)

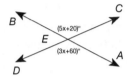

A a) Since $\overleftrightarrow{AB}$ and $\overleftrightarrow{CD}$ are straight lines intersecting at point E, $\angle BEC$ and $\angle AED$ are, by definition, vertical angles. As such, they are congruent and their measures are equal. Therefore, if we let x represent the fixed quantity, we can set up the following equality, and solve for the unknown quantity.

$$5x + 20° = 3x + 60°$$

$$5x - 3x = 60° - 20°$$

$$2x = 40°$$

$$x = 20°$$

Therefore, the value of the unknown quantity is 20°.

b) From the information given about $\angle BEC$, we know that $m\angle BEC = 5x + 20°$. By substitution, we have

$$m\angle BEC = 5(20°) + 20° = 100° + 20° = 120°.$$

Therefore, the measure of $\angle BEC$ is 120°.

c) We know that $\overleftrightarrow{AB}$ is a straight line; therefore, $\angle CEA$ is the supplement of $\angle BEC$. Since the sum of the measure of two supplements is 180°, the following calculation can be made:

$$m\sphericalangle CEA + m\sphericalangle BEC \quad = \quad 180°$$

$$m\sphericalangle CEA \quad = \quad 180° - m\sphericalangle BEC,$$

Substituting in our value for $m \sphericalangle BEC$, we obtain:

$$m\sphericalangle CEA \quad = \quad 180° - 120° = 60°$$

Therefore, the measure of $\sphericalangle CEA$ is 60°.

Theorem 8

Any point on the perpendicular bisector of a given line segment is equidistant from the ends of the segment.

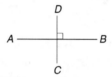

Theorem 9

If a point is equidistant from the ends of a line segment, this point must lie on the perpendicular bisector of the segment.

Theorem 10

If two points are equidistant from the ends of a line segment, these points determine the perpendicular bisector of the segment.

Theorem 11

Every line segment has exactly one midpoint.

Theorem 12

There exists one and only one perpendicular to a line through a point outside the line. Take point C outside line $\overset{\leftrightarrow}{AB}$. $\overset{\leftrightarrow}{OC}$ represents the only perpendicular to $\overset{\leftrightarrow}{AB}$ which passes through C.

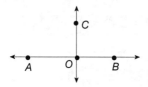

Theorem 13

If the exterior sides of adjacent angles are perpendicular to each other, then the adjacent angles are complementary.

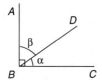

Theorem 14

Adjacent angles are supplementary if their exterior sides form a straight line.

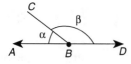

Theorem 15

Two angles which are equal and supplementary to each other are right angles.

CHAPTER 3

Congruent Angles

and Congruent

Line Segments

3.1 Definitions

Definition 1

Two or more geometric figures are congruent when they have the same shape and size. The symbol for congruence is ≅; hence, if triangle *ABC* is congruent to triangle *DEF*, we write $\triangle ABC \cong \triangle DEF$.

Definition 2

Two line segments are congruent if and only if they have the same measure.

Note: The expression "if and only if" can be used any time both a statement and the converse of that statement are true. Using definition 2, we can rewrite the statement as "two line segments have the same measure if and only if they are congruent." The two statements are identical.

Definition 3

Two angles are congruent if and only if they have the same measure.

Problem Solving Example:

 In the figure shown, $\triangle ABC$ is an isosceles triangle, such that $\overline{BA} \cong \overline{BC}$. Line segment $\overline{AD}$ bisects $\measuredangle BAC$ and $\overline{CD}$ bisects $\measuredangle BCA$. Prove that $\triangle ADC$ is an isosceles triangle.

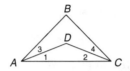

 In order to prove $\triangle ADC$ is isosceles, we must prove that 2 of its sides, $\overline{AD}$ and $\overline{CD}$, are congruent. To prove $\overline{AD} \cong \overline{CD}$ in $\triangle ADC$, we have to prove that the angles opposite $\overline{AD}$ and $\overline{CD}$, $\measuredangle 1$ and $\measuredangle 2$, are congruent.

Statement	Reason
1. $\overline{BA} \cong \overline{BC}$	1. Given.
2. $\measuredangle BAC \cong \measuredangle BCA$ or $m\measuredangle BAC = m\measuredangle BCA$	2. If two sides of a triangle are congruent, then the angles opposite them are congruent.
3. $\overline{AD}$ bisects $\measuredangle BAC$ $\overline{CD}$ bisects $\measuredangle BCA$	3. Given.
4. $m\measuredangle 1 = 1/2 m\measuredangle BAC$ $m\measuredangle 2 = 1/2 m\measuredangle BCA$	4. The bisector of an angle divides the angle into two angles whose measures are equal.
5. $m\measuredangle 1 = m\measuredangle 2$	5. Halves of equal quantities are equal.
6. $\measuredangle 1 \cong \measuredangle 2$	6. If the measure of two angles are equal, then the angles are congruent.
7. $\overline{CD} \cong \overline{AD}$	7. If two angles of a triangle are congruent, then the sides opposite these angles are congruent.
8. $\triangle ADC$ is an isosceles triangle.	8. If a triangle has two congruent sides, then it is an isosceles triangle.

3.2   Theorems

Theorem 1

Every line segment is congruent to itself.

Theorem 2

Every angle is congruent to itself.

Let R be a relation on a set A. Then:

R is reflexive if aRa for every a in A.

R is symmetric if aRb implies bRa.

R is anti-symmetric if aRb and bRa imply $a = b$.

R is transitive if aRb and bRc imply aRc.

Note: The term aRa means the relation R performed on a yields a. The term aRb means the relation R performed on a yields b.

3.3   Postulates

By definition, a relation R is called an equivalence relation if relation R is reflexive, symmetric, and transitive.

Postulate 1

Congruence of segments is an equivalence relation.

(1) Congruence of segments is reflexive.

If $\overline{AB} \cong \overline{AB}$, $\overline{AB}$ is congruent to itself.

(2) Congruence of segments is symmetric.

If $\overline{AB} \cong \overline{CD}$, then $\overline{CD} \cong \overline{AB}$.

(3) Congruence of segments is transitive.

If $\overline{AB} \cong \overline{CD}$ and $\overline{CD} \cong \overline{EF}$, then $\overline{AB} \cong \overline{EF}$.

Postulate 2

Congruence of angles is an equivalence relation, i.e., reflexive, symmetric, and transitive.

Postulate 3

Any geometric figure is congruent to itself.

Postulate 4

A congruence may be reversed.

Postulate 5

Two geometric figures congruent to the same geometric figure are congruent to each other.

3.4   Theorems

Theorem 3

Given a line segment $\overline{AB}$ and a ray $\overrightarrow{XY}$, there exists one and only one point O on $\overrightarrow{XY}$ such that $\overline{AB} \cong \overline{XO}$.

Theorem 4

If $\overline{AB} = \overline{CD}$, Q bisects $\overline{AB}$ and P bisects $\overline{CD}$, then $\overline{AQ} \cong \overline{CP}$, $\overline{AQ} = \overline{CP}$.

Theorem 5

If $m \angle ABC = m \angle DEF$, and $\overrightarrow{BX}$ and $\overrightarrow{EY}$ bisect $\angle ABC$ and $\angle DEF$, respectively, then $m \angle ABX = m \angle DEY$.

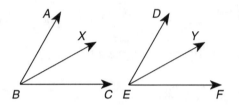

Theorem 6

Let P be in the interior of $\angle ABC$ and Q be in the interior of $\angle DEF$. If $m \angle ABP = m \angle DEQ$ and $m \angle PBC = QEF$, then $m \angle ABC = m \angle DEF$.

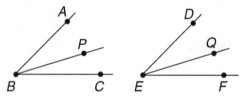

Theorem 7

Let P be in the interior of $\angle XYZ$ and Q be in the interior of $\angle ABC$. If $m \angle XYZ = m \angle ABC$ and $m \angle XYP = m \angle ABQ$ then $m \angle PYZ = m \angle QBC$.

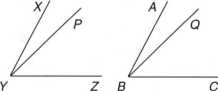

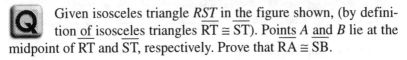

Problem Solving Example:

 Given isosceles triangle *RST* in the figure shown, (by definition of isosceles triangles $\overline{RT} \cong \overline{ST}$). Points *A* and *B* lie at the midpoint of $\overline{RT}$ and $\overline{ST}$, respectively. Prove that $\overline{RA} \cong \overline{SB}$.

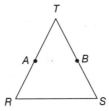

 This solution is best presented as a formal proof.

Statement	Reason
1. $\overline{RT} \cong \overline{ST}$ or $RT = ST$	1. Given.
2. *A* is the midpoint of $\overline{RT}$	2. Given.
3. $RA = 1/2\ RT$	3. The midpoint of a line segment divides the line segment into two equal halves.
4. *B* is the midpoint of $\overline{ST}$	4. Given.
5. $\overline{SB} = 1/2\ \overline{TS}$	5. The midpoint of a line segment divides the line segment into two equal halves.
6. $RA = SB$	6. Division Postulate: Halves of equal quantities are equal. Statements 3 and 5.
7. $RA \cong SB$	7. If two line segments are of equal length, then they are congruent.

Quiz: Method of Proof –
Congruent Angles & Line Segments

Refer to the diagram and find the appropriate solution.

1. Find *a*.

    (A) 38°

    (B) 68°

    (C) 78°

    (D) 90°

    (E) 112°

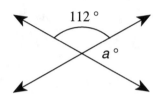

2. Find *c*.

    (A) 32°

    (B) 48°

    (C) 58°

    (D) 82°

    (E) 148°

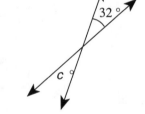

3. Determine *x*.

    (A) 21°

    (B) 23°

    (C) 51°

    (D) 102°

    (E) 153°

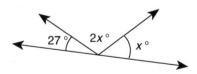

4. Find *z*.

    (A) 29°

    (B) 54°

    (C) 61°

    (D) 88°

    (E) 92°

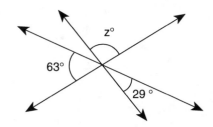

5. In the figure shown, if $\overline{BD}$ is the bisector of angle *ABC*, and angle *ABD* is one-fourth the size of angle *XYZ*, what is the size of angle *ABC*?

    (A) 21°

    (B) 28°

    (C) 42°

    (D) 63°

    (E) 168°

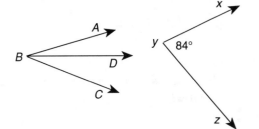

6. $\overrightarrow{BA} \perp \overrightarrow{BC}$ and $m \angle DBC = 53°$.
    Find m∡*ABD*.

    (A) 27°

    (B) 33°

    (C) 37°

    (D) 53°

    (E) 90°

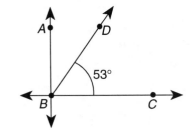

7.   If $n \perp p$, which of the following statements is true?

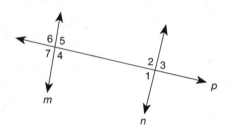

    (A) $\angle 1 \cong \angle 2$

    (B) $\angle 4 \cong \angle 5$

    (C) $m\angle 4 + m\angle 5 > m\angle 1 + m\angle 2$

    (D) $m\angle 3 > m\angle 2$

    (E) $m\angle 4 = 90°$

8.   In the figure, $p \perp t$ and $q \perp t$, which of the following statements is false?

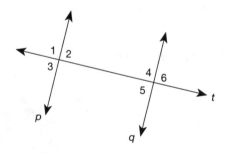

    (A) $\angle 1 \cong \angle 4$

    (B) $\angle 2 \cong \angle 3$

    (C) $m\angle 2 + m\angle 3 = m\angle 4 + m\angle 6$

    (D) $m\angle 5 + m\angle 6 = 180°$

    (E) $m\angle 2 > m\angle 5$

9. If $a \parallel b$, find z.

    (A) 26°

    (B) 32°

    (C) 64°

    (D) 86°

    (E) 116°

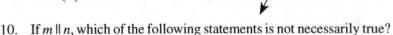

10. If $m \parallel n$, which of the following statements is not necessarily true?

    (A) $\angle 2 \cong \angle 5$

    (B) $\angle 3 \cong \angle 6$

    (C) $m\angle 4 + m\angle 5 = 180°$

    (D) $\angle 1 \cong \angle 6$

    (E) $m\angle 7 + m\angle 3 = 180°$

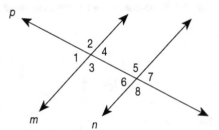

ANSWER KEY

1.	(B)	6.	(C)
2.	(A)	7.	(A)
3.	(C)	8.	(E)
4.	(D)	9.	(C)
5.	(C)	10.	(B)

CHAPTER 4

Triangles and Congruent Triangles

4.1 Triangles

Definition 1

A closed three-sided geometric figure is called a triangle. The points of the intersection of the sides of a triangle are called the vertices of the triangle.

Definition 2

A side of a triangle is a line segment whose endpoints are the vertices of two angles of the triangle.

Definition 3

A triangle with no equal sides is called a scalene triangle.

Definition 4

A triangle having two equal sides is called an isosceles triangle. The third side is called the base of the triangle. The angle opposite the base is called the vertex angle. ⊀A is the vertex angle.

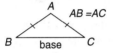

Definition 5

The perimeter of a triangle is the sum of the measures of the sides of the triangle.

Definition 6

An interior angle of a triangle is an angle formed by two sides and includes the third side within its collection of points.

Problem Solving Examples:

Prove that a scalene triangle has no two angles congruent.

We shall prove the given theorem by the method of "proof by contradiction." This method entails proving that the contrapositive of the given theorem is true. If this is so, then the original theorem is true.

The contrapositive of the given theorem is: If two angles of a triangle are congruent, then it is not scalene. If two angles of a triangle are congruent, it is isosceles. This implies that two sides of the triangle are congruent. But, by definition, a scalene triangle has no sides congruent. Hence, if two angles of a triangle are congruent, it is not scalene. We have therefore proven the contrapositive of the given theorem to be true; hence, the given theorem is true.

 Prove that if a triangle has no two angles congruent, then it is scalene.

We prove this theorem by the method of contradiction. That is, we prove that the contrapositive of the given theorem is true. It then follows that the theorem itself is true.

The contrapositive of the given theorem is: If a triangle is non-scalene, then at least two of its angles are congruent.

By definition, a scalene triangle has no two sides congruent. Hence, a non-scalene triangle has at least two sides congruent. This means that a non-scalene triangle is either an isosceles triangle or an equilateral triangle. Each of the latter triangles has at least two angles congruent. Hence, we have shown that a non-scalene triangle has at least two angles congruent. Since the contrapositive of the given theorem is true, the given theorem is true.

Definition 7

An equilateral triangle is a triangle having three equal sides.

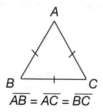

$$\overline{AB} = \overline{AC} = \overline{BC}$$

Definition 8

A triangle with one obtuse angle (greater than 90°) is called an obtuse triangle.

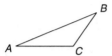

Definition 9

An acute triangle is a triangle with three acute angles (less than 90°).

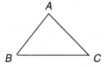

Definition 10

A triangle with a right angle is called a right triangle. The side opposite the right angle in a right triangle is called the hypotenuse of the right triangle. The other two sides are called legs of the right triangle.

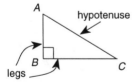

Definition 11

An altitude of a triangle is a line segment from a vertex of the triangle perpendicular to the opposite side.

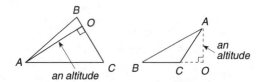

Definition 12

A line segment connecting a vertex of a triangle and the midpoint of the opposite side is called a median of the triangle.

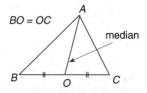

Problem Solving Examples:

 Prove that an equilateral triangle has three equal angles.

 Draw equilateral $\triangle ABC$.

Hence, the problem can be restated as:

Given: equilateral $\triangle ABC$

Prove: $m \angle A = m \angle B = m \angle C$.

Statement	Reason
1. equilateral $\triangle ABC$	1. Given.
2. $AC \cong BC \cong AB$	2. Definition of an equilateral triangle.
3. $\angle A \cong \angle B$ $\angle A \cong \angle C$ $\angle B \cong \angle C$	3. If two sides of a triangle are congruent, then the angles opposite those sides are congruent.
4. $\angle A \cong \angle B \cong \angle C$	4. Transitive Property of Congruence.

Q As seen in the accompanying diagram, $\triangle ABC$ is constructed in such a way that the measure of $\angle A$ equals $9x$, $m\angle B$ equals $3x - 6$, and $m\angle C$ equals $11x + 2$, x being some unknown. Show that $\triangle ABC$ is a right triangle.

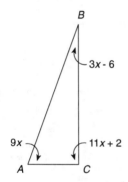

A A triangle is a right triangle if one of its angles is a right angle. The best way to determine the "rightness" of this triangle would be to sum the measures of all its angles, set this sum equal to 180°, and solve for the unknown x. If the measure of one angle turns out to be 90°, then it is a right angle and the triangle is a right triangle. The algebra is as follows:

$$m\angle A + m\angle B + m\angle C = 180$$
$$9x + 3x - 6 + 11x + 2 = 180$$
$$23x - 4 = 180$$
$$23x = 184$$
$$x = 8$$

Therefore,

∡A measures (9)(8) or 72

∡B measures (3)(8) – 6 or 18

and ∡C measures (11)(8) + 2 or 90

Therefore, since ∡C measures 90°, Δ*ABC* is a right triangle.

Definition 13

A line that bisects and is perpendicular to a side of a triangle is called a perpendicular bisector of that side.

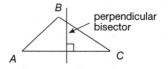

Definition 14

An angle bisector of a triangle is a line that bisects an angle and extends to the opposite side of the triangle.

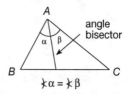

Definition 15

The line segment that joins the midpoints of two sides of a triangle is called a midline of the triangle.

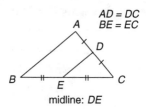

Definition 16

An exterior angle of a triangle is an angle formed outside a triangle by one side of the triangle and the extension of an adjacent side.

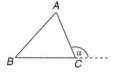

Definition 17

A triangle whose three interior angles have equal measures is said to be equiangular.

Definition 18

Three or more lines (or rays or segments) are concurrent if there exists one point common to all of them, that is, if they all intersect at the same point.

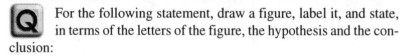

Problem Solving Examples:

Q For the following statement, draw a figure, label it, and state, in terms of the letters of the figure, the hypothesis and the conclusion:

If the bisector of the vertex angle of an isosceles triangle is drawn, then the bisector is perpendicular to the base of the triangle.

A The hypothesis gives the specific guidelines for the diagram to be drawn with one exception. This exception is filled in by the conclusion of the statement.

The hypothesis is "the bisector of the vertex angle of an isosceles triangle is drawn." Draw an isosceles triangle and then the bisector of its vertex angle. The bisector will intersect the base of the triangle, but in what specific manner is not yet known.

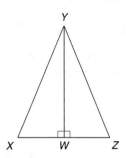

The conclusion tells us that the bisector will be perpendicular to the base of the triangle.

In terms of the diagram, the hypothesis can be written as: "If $\overline{YW}$ is the bisector of $\angle Y$ in $\triangle XYZ$,"

Conclusion: "then $\overline{YW} \perp \overline{XY}$."

 In $\triangle ABC$, $\overline{AC} \cong \overline{BC}$. The measure of an exterior angle of vertex C is represented by $5x + 10$. If $\angle A$ measures $30°$, find the value of x.

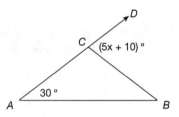

 To solve this problem, we relate $m\angle DCB$ to $\angle A$ and $\angle B$. First, since $\overline{AC} \cong \overline{BC}$, we know that $\triangle ACB$ is isosceles. Because the base angles of an isosceles triangle are congruent, $m\angle A = m\angle B = 30°$.

Secondly, $\angle DCB$ forms a linear pair with one of the interior angles of $\triangle ACB$. Thus $\angle DCB$ is an exterior angle and must be equal in measure to the sum of the remote interior angles, $\angle A$ and $\angle B$.

(I) $\qquad m\angle DCB = m\angle A + m\angle B$

(II) $\qquad 5x + 10 = 30 + 30$

(III) $\qquad 5x = 50°$

(IV) $\qquad x = 10°.$

Theorem 1

The three lines containing the altitudes of a triangle are concurrent.

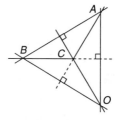

Theorem 2

The medians of a triangle are concurrent at a point which is two-thirds the distance from any vertex to the midpoint of the opposite side.

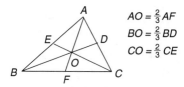

$$AO = \tfrac{2}{3}AF$$
$$BO = \tfrac{2}{3}BD$$
$$CO = \tfrac{2}{3}CE$$

Point O is called the centroid of $\triangle ABC$.

Theorem 3

The perpendicular bisectors of the sides of a triangle are concurrent at a point that is equidistant from any vertex of the triangle.

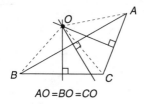

$$AO = BO = CO$$

Theorem 4

The angle bisectors of a triangle are concurrent at a point which is equidistant from any side of the triangle.

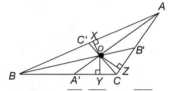

Angle bisectors $\overline{AA'}$, $\overline{BB'}$, and $\overline{CC'}$ meet at point O, and $OX = OY = OZ$.

Theorem 5

The measure of an exterior angle of a triangle is equal to the sum of the measures of the two non-adjacent interior angles of that triangle.

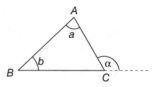

$$m \sphericalangle \alpha = m \sphericalangle a + m \sphericalangle b$$

Problem Solving Examples:

Show that the lines containing the three altitudes of a triangle are concurrent.

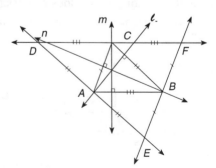

A One theorem states that the perpendicular bisectors are concurrent. Here we are asked to show that another set of lines are concurrent. If we show that the lines that contain the altitudes of a given triangle also contain the perpendicular bisectors of a second triangle, then the concurrency follows immediately.

Given: The attitudes of $\triangle ABC$, ℓ, m, and n.

Prove: ℓ, m, n are concurrent.

Statement	Reason
1. The altitudes of $\triangle ABC$, ℓ, m, and n	1. Given.
2. Construct $\overleftrightarrow{DF} \parallel \overline{AB}$ through C, $\overleftrightarrow{BE} \parallel \overline{BC}$ through A, and $\overleftrightarrow{EF} \parallel \overleftrightarrow{AC}$ through B	2. Through a given external point only one line can be drawn parallel to a given line.
3. Quadrilateral $ABFC$ is a parallelogram	3. If opposite sides of a quadrilateral ($\overline{AC}$ and $\overline{BF}$, $\overline{CF}$ and $\overline{AB}$) are parallel, then the quadrilateral is a parallelogram.
4. Quadrilateral $ACBE$ is a parallelogram Quadrilateral $ADCB$ is a parallelogram	4. Same reason as Step 3.
5. (a) $\overline{DC} \cong \overline{AB}$, $\overline{CF} \cong \overline{AB}$ (b) $\overline{BE} \cong \overline{CA}$, $\overline{FB} \cong \overline{CA}$ (c) $\overline{DA} \cong \overline{CB}$, $\overline{AE} \cong \overline{CB}$	5. Opposite sides of a parallelogram are congruent.
6. (a) $\overline{DC} \cong \overline{CF}$ (b) $\overline{BE} \cong \overline{FB}$ (c) $\overline{DA} \cong \overline{AE}$	6. Segments congruent to the same segments are congruent.
7. (a) C is the midpoint of $\overline{DF}$ (b) B is the midpoint of $\overline{FE}$ (c) A is the midpoint of $\overline{DE}$	7. Definition of midpoint.
8. (a) $m \perp \overline{AB}$ (b) $n \perp \overline{AC}$ (c) $\ell \perp \overline{CB}$	8. Definition of the altitude of a triangle.

9. (a) $m \perp \overline{DF}$
    (b) $n \perp \overline{FE}$
    (c) $\ell \perp \overline{DE}$

9. In a plane, if a line is perpendicular to one of two parallel lines, then it is perpendicular to the other parallel line.

10. (a) m is the perpendicular bisector of $\overline{DF}$
    (b) n is the perpendicular bisector of $\overline{FE}$
    (c) ℓ is the perpendicular bisector of $\overline{DE}$

10. A line perpendicular to a segment at its midpoint is the perpendicular bisector of the segment.

11. $\overline{DF}, \overline{FE}, \overline{DE}$ are sides of $\triangle DFE$

11. In a plane, three non-colinear points determine a triangle.

12. n, m, and ℓ are concurrent

12. The perpendicular bisectors of the sides of a triangle are concurrent.

 Show that the angle bisectors of a triangle are concurrent at a point equidistant from the sides of the triangle.

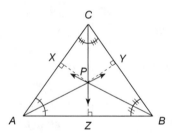

 The angle bisector of any angle is the set of points in the interior of the angle that are equidistant from the sides of the angle.

In $\triangle ABC$, the intersection of the bisector of $\angle A$ and the bisector of $\angle B$ is a point P. Because P is on the bisector of $\angle A$, the distance from P to $\overline{AC}$ equals the distance from P to $\overline{AB}$. Because P is on the bisector of $\angle B$, the distance from P to $\overline{BC}$ equals the distance from P to $\overline{AB}$. By transitivity, the distance from P to $\overline{AC}$ equals the distance from P to $\overline{BC}$, i.e. P is equidistant from the sides of $\angle C$. Therefore, point P is also on the bisector of $\angle C$, and the three bisectors are concurrent at a point equidistant from the sides of the triangle.

Given: The angle bisectors of $\triangle ABC$.

Prove: The bisectors are concurrent at a point equidistant from the sides.

Statement	Reason
1. The angle bisectors of $\triangle ABC$	1. Given.
2. Let P be the point of intersection of the bisectors of $\angle A$ and $\angle B$.	2. In a plane, two non-parallel, non-coincident lines intersect in a unique point.
3. P is in the interior of $\angle A$ P is in the interior of $\angle B$	3. All points (except for the vertex) of the angle bisector lie in the interior of the angle.
4. P is in the interior of $\triangle ABC$	4. If a point is in the interior of the two angles of a triangle, then it is in the interior of the triangle.
5. Let X, Y, and Z be the points on sides $\overline{AC}$, $\overline{CB}$, and $\overline{AB}$ such that $\overline{PX} \perp \overline{AC}$, $\overline{PY} \perp \overline{CB}$, and $\overline{PZ} \perp \overline{AB}$	5. From a given external point, only one line can be drawn perpendicular to a given line.
6. $\overline{PX}$, $\overline{PY}$, and $\overline{PZ}$ are the distances from P to the sides	6. The distance from an external point to a line is the length of the perpendicular segment from the point to the line.
7. $\overline{PX} = \overline{PZ}$ $\overline{PY} = \overline{PZ}$	7. All points on the angle bisector are equidistant from the sides of the angle.
8. $\overline{PX} = \overline{PY}$	8. Transitivity Postulate.
9. P is in the interior of $\angle C$	9. All points in the interior of a triangle are in the interior of each of the angles.

10. *P* is on the angle bisector of ∢*C*	10. All points in the interior of an angle that are equidistant from the sides of the angle are on the angle bisector.
11. *P* is equidistant from the sides of Δ*ABC*	11. Follows from Steps 7 and 8.
12. The angle bisectors are concurrent at a point equidistant from the sides	12. Lines are concurrent if their intersection consists of at least one point. Also, Step 11.

Theorem 6

Every angle of a triangle has one and only one bisector.

Theorem 7

The midline of a triangle is parallel to the third side of the triangle.

Theorem 8

The midline of a triangle is half as long as the third side of the triangle.

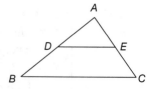

DE is a midline of Δ*ABC* such that
AD = *BD* and *AE* = *CE*. Therefore,
DE = 1/2 *BC*.

Theorem 9

The sum of the measures of the interior angles of a triangle is 180°.

Theorem 10

If two angles of one triangle are equal respectively to two angles of a second triangle, their third angles are equal.

Problem Solving Example:

Given: *DA* bisects ∢*CAB*. *DB* bisects ∢*CBA*; $m∢1 = m∢2$.
Prove: $CA = CB$.

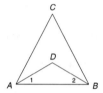

 This proof will involve using the definition of angle bisector and the substitution and addition postulates to prove $m∢$ CAB = ∢ CBA. The desired results will be obtained by recalling that, in any triangle, the sides that lie opposite angles of equal measure are of equal length.

Statement	Reason
1. *DA* bisects ∢*CAB* *DB* bisects ∢*CBA*	1. Given.
2. $m∢CAD = m∢1$ $m∢CBD = m∢2$	2. An angle bisector divides an angle into two angles of equal measure.
3. $m∢1 = m∢2$	3. Given.
4. $m∢CBD = m∢1$	4. A quantity may be substituted for its equal.
5. $m∢CAD = m∢CBD$	5. If two quantities are equal to the same quantity, or equal quantities, then they are equal to each other.
6. $m∢CAD + m∢1 =$ $m∢CBD + m∢2$	6. If equal quantities are added to equal quantities, the sums are equal quantities.

7. $m \angle CAB = m \angle CAD + m \angle 1$ $m \angle CBA = m \angle CBD + m \angle 2$	7. The measure of the whole is equal to the sum of the measures of its parts.
8. $m \angle CAB = m \angle CBA$	8. If two quantities are equal to the same quantity, or equal quantities, then they're equal to each other.
9. $CA = CB$	9. If a triangle has two angles of equal measure then the sides opposite those angles are equal in length.

Theorem 11

A triangle can have at most one right or obtuse angle.

Theorem 12

If a triangle has two equal angles, then the sides opposite those angles are equal.

Theorem 13

If two sides of a triangle are equal, then the angles opposite those sides are equal.

Theorem 14

The sum of the exterior angles of a triangle, taking one angle at each vertex, is 360°.

Theorem 15

A line that bisects one side of a triangle and is parallel to a second side, bisects the third side.

Theorem 16 (The Law of Cosines)

For a given triangle $\triangle ABC$ with sides of length a, b, and c,

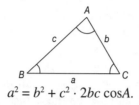

$$a^2 = b^2 + c^2 \cdot 2bc \cos A.$$

Problem Solving Examples:

Given: $m \sphericalangle A = m \sphericalangle B$. $AD = BE$.
Prove: $m \sphericalangle CDE = m \sphericalangle CED$.

   This proof will revolve around the theorem stating that two angles of a triangle are of equal measure if and only if the sides opposite them are of equal length.

Statement	Reason
1. $m \sphericalangle A = m \sphericalangle B$	1. Given.
2. $CA = CB$	2. If a triangle has two angles of equal measure, then the sides opposite those angles are equal in length.
3. $CB = CE + BE$ $CA = CD + AD$	3. The measure of the whole is equal to the sum of the measures of its parts.
4. $CD + AD =$ $CE + BE$	4. A quantity may be substituted for its equal.
5. $AD = BE$	5. Given.
6. $CD = CE$	6. If equal quantities are subtracted from equal quantities, the differences are equal quantities.
7. $m \sphericalangle CDE =$ $m \sphericalangle CED$	7. If two sides of a triangle are equal, then the angles opposite those sides are equal.

Prove that a triangle can have, at most, one obtuse angle.

 We prove this result indirectly. We assume that a triangle can have more than one obtuse angle. Then we show this leads to a contradiction.

Assume that there exists a $\triangle ABC$ with more than one, say two, obtuse angles. Let $\angle A$ and $\angle B$ be the obtuse angles. Since the measure of an obtuse angle must be greater than 90° but less than 180°, $m\angle A > 90$ and $m\angle B > 90$. Thus, by the Addition Property of Inequality, we have

  (i) $m\angle A + m\angle B > 180$.

Since the measure of $\angle C$ must be greater than zero, that is, $m\angle C > 0$, it follows that

  (ii) $m\angle A + m\angle B + m\angle C > 180$.

This contradicts the theorem that states that the sum of the measures of the angles of a triangle must equal 180°. Therefore, our assumption is false, and a triangle has, at most, one obtuse angle.

 Given: $AC = BC$. $AD = BD$.
Prove: $m\angle CAD = m\angle CBD$.

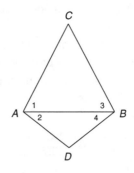

The required results can be derived by proving that the angles that make up ∡CAD and ∡CBD are equal to each other.

Statement	Reason
1. $AC = BC$ $AD = BD$	1. Given.
2. $m∡2 = m∡4$ $m∡1 = m∡3$	2. If two sides of a triangle are of equal length, then the angles opposite those sides have equal measure.
3. $m∡1 + m∡2 = m∡3 + m∡4$	3. If equal quantities are added to equal quantities, the sums are equal quantities.
4. $m∡CAD = m∡1 + m∡2$ $m∡CBD = m∡3 + m∡4$	4. The measure of the whole is equal to the sum of the measures of its parts.
5. $m∡CAD = m∡CBD$	5. A quantity may be substituted for its equal.

4.2 Isosceles, Equilateral, and Right Triangles

Theorem 1

The length of the median to the hypotenuse of a right triangle is equal to one-half the length of the hypotenuse.

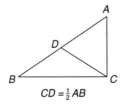

$CD = \frac{1}{2}AB$

Theorem 2

In a right triangle, the square of the hypotenuse is equal to the sum of the squares of the other two sides. This is commonly known as the theorem of Pythagoras or the Pythagorean theorem.

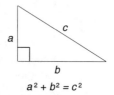

$$a^2 + b^2 = c^2$$

Theorem 3

If a triangle has sides of length a, b and c, and $c^2 = a^2 + b^2$, then the triangle is a right triangle.

Theorem 4

In a 30°– 60° right triangle, the hypotenuse is twice the length of the side opposite the 30° angle. The side opposite the 60° angle is equal to the length of the side opposite the 30° angle multiplied by $\sqrt{3}$.

In an isosceles 45° right triangle, the hypotenuse is equal to the length of one of its legs multiplied by $\sqrt{2}$.

(1) $AB = 2AC$;   (2) $BC = AC\sqrt{3}$;   and (3) $XY = XZ\sqrt{2} = YZ\sqrt{2}$

By rearranging these expressions, we can obtain:

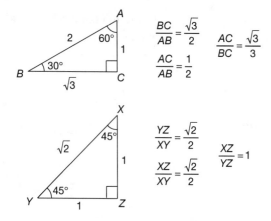

$$\frac{BC}{AB} = \frac{\sqrt{3}}{2} \qquad \frac{AC}{BC} = \frac{\sqrt{3}}{3}$$

$$\frac{AC}{AB} = \frac{1}{2}$$

$$\frac{YZ}{XY} = \frac{\sqrt{2}}{2} \qquad \frac{XZ}{YZ} = 1$$

$$\frac{XZ}{XY} = \frac{\sqrt{2}}{2}$$

Theorem 5

The altitude of an equilateral triangle equals $\frac{\sqrt{3}}{2}$ times the measure of a side of the triangle.

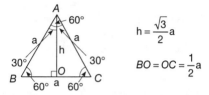

$$h = \frac{\sqrt{3}}{2}a$$

$$BO = OC = \frac{1}{2}a$$

Theorem 6

The base angles of an isosceles triangle are congruent, where the two sides adjacent to the base are equal.

Problem Solving Examples:

 The length of the median drawn to the hypotenuse of a right triangle is 12 inches. Find the length of the hypotenuse. (see figure)

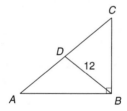

 A theorem tells us that the length of the median to the hypotenuse of a right triangle is equal to one-half the length of the hypotenuse. We must identify the median, the hypotenuse, their respective lengths, and substitute them according to the rule cited, and solve for any unknowns.

$\overline{AC}$ is the hypotenuse, $\overline{BD}$ is the median of length 12", and the length of $\overline{AC}$ is unknown. By applying the above theorem we know,

(1) $\overline{BD} = \frac{1}{2}\overline{AC}$ and, by, substitution

(2) 12" $= \frac{1}{2}\overline{AC}$ which implies that $\overline{AC} = 24$".

Therefore, the length of hypotenuse $\overline{AC}$ is 24 in.

 Is a triangle with sides 3, 7, and 11 inches a right triangle?

A Recall that the converse of the Pythagorean Theorem is also true. It states that if a triangle has sides of length a, b, and c and $c^2 = a^2 + b^2$, then the triangle is a right triangle. Let $a = 3$, $b = 7$, and $c = 11$. We compute the squares:

$$
\begin{aligned}
a^2 &= 3^2 &= 9 \\
b^2 &= 7^2 &= 49 \\
c^2 &= 11^2 &= 121 \\
121 &\neq 49 + 9.
\end{aligned}
$$

Since the sum of any two of these squares does not equal the third square, the triangle is not a right triangle.

Corollary 1

The angle bisectors of the base angles of an isosceles triangle are congruent.

Corollary 2

The bisector of the vertex angle of an isosceles triangle is also the perpendicular bisector of the base of the triangle.

Corollary 3

All equiangular triangles are also equilateral and every equilateral triangle is equiangular.

Corollary 4

The three angles of an equilateral triangle each have a measure of 60°.

Corollary 5

The acute angles of a right triangle are complementary.

Corollary 6

The two acute angles of an isosceles right triangle each have a measure of 45°.

Problem Solving Examples:

 Prove that the base angles of an isosceles right triangle have measure 45°.

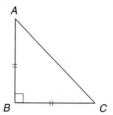

 As drawn in the figure, $\triangle ABC$ is an isosceles right triangle with base angles BAC and BCA. The sum of the measures of the angles of any triangle is 180°. For $\triangle ABC$, this means

(1) $m\sphericalangle BAC + m\sphericalangle BCA + m\sphericalangle ABC = 180°$.

But $m\sphericalangle ABC = 90°$ because ABC is a right triangle. Furthermore, $m\sphericalangle BCA = m\sphericalangle BAC$, since the base angles of an isosceles triangle are congruent. Using these facts in equation (1)

$$m\sphericalangle BAC + m\sphericalangle BCA + 90° = 180°$$

or $2m\angle BAC = 2m\angle BCA = 90°$

or $m\angle BAC = m\angle BCA = 45°$.

Therefore, the base angles of an isosceles right triangle have measure 45°.

 Prove that each angle of an equilateral triangle has measure 60°.

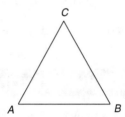

 We will apply the angle sum theorem to an equilateral triangle to prove that the measure of each angle is 60°.

Let the equilateral triangle be $\triangle ABC$. By the angle sum theorem,

(1) $m\angle A + m\angle B + m\angle C = 180°$.

But, the angles of an equilateral triangle are congruent. That is,

$$\angle A \cong \angle B \cong \angle C.$$

Thus,

(2) $m\angle A = m\angle B = m\angle C$.

Substituting (2) into (1),

$$3m\angle A = 180°$$

or $m\angle A = 60°$.

By equation (2), all the angles of an equilateral triangle have measure 60°.

4.3   Congruent Triangles

Definition

Two polygons are congruent if there is a one-to-one correspondence between their vertices such that all pairs of corresponding sides have equal measures and all pairs of corresponding angles have equal measures. This is denoted by $\cong$.

Theorem 1

Triangle congruence is an equivalence relation.

By definition, a relation R is called an equivalence relation if relation R is reflexive, symmetric, and transitive.

Properties of Congruence:

A) Reflexive property: $\triangle ABC \cong \triangle ABC$.

B) Symmetric property: If $\triangle ABC \cong \triangle DEF$, then $\triangle DEF \cong \triangle ABC$

C) Transitive Property: If $\triangle ABC \cong \triangle DEF$, and $\triangle DEF \cong \triangle RST$, then $\triangle ABC \cong \triangle RST$.

Problem Solving Example:

Q The vertices of $\triangle ABC$, when drawn on the Cartesian plane, are $A(-3, 0)$, $B(3, 0)$, and $C(0, 2)$. Prove that $\triangle ABC$ is an isosceles triangle.

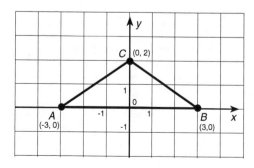

A If we let the origin be point *O*, and prove $\triangle COA \cong \triangle COB$, then, by corresponding parts, we can conclude $\overline{CA} \cong \overline{CB}$. This will be sufficient to show $\triangle ABC$ is isosceles, since an isosceles triangle is defined to be one in which two sides are congruent. The SAS $\cong$ SAS method will be used.

Since, by definition of the Cartesian plane, the *y*-axis $\perp$ *x*-axis; thus $\angle COA$ and $\angle COB$ are right angles and, they are congruent.

OA = 3 units and *OB* = 3 units and, because their lengths are equal, they are, therefore, congruent.

We now have congruence between one angle in each triangle and one corresponding adjacent side. The other adjacent side, $\overline{OC}$, is common to both triangles and, by reflexivity of congruence, is congruent to itself.

Therefore, $\triangle COA \cong \triangle COB$ by SAS $\cong$ SAS

Thus, $\overline{CA} \cong \overline{CB}$, because corresponding sides of congruent triangles are congruent.

Therefore, $\triangle ABC$ is isosceles because it is a triangle which has two congruent sides.

Postulate 1

If three sides of one triangle are equal, respectively, to three sides of a second triangle, the triangles are congruent. (SSS = SSS)

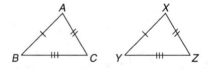

Postulate 2

If two sides and the included angle of one triangle are equal, respectively, to two sides and the included angle of a second triangle, the triangles are congruent. (SAS = SAS)

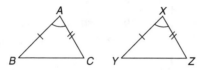

Postulate 3

If two angles and the included side of one triangle are equal, respectively, to two angles and the included side of a second triangle, the triangles are congruent. (ASA = ASA)

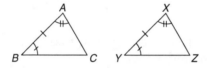

Problem Solving Examples:

 Given: $\triangle ABC$. $DC = EC$. $\angle 1 = \angle 2$.
Prove: $\overline{AC} \cong \overline{BC}$.

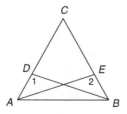

The segments in question, $\overline{AC}$ and $\overline{BC}$, are corresponding parts of overlapping Δ's *CDB* and *CEA*. The proof is set up using the ASA postulate to show Δ*CDB* ≅ Δ*CEA*.

Statement	Reason
1. ⊀1 = ⊀2	1. Given.
2. ⊀1 and ⊀ *CDB* are supplements ⊀2 and ⊀ *CEA* are supplements	2. Adjacent angles whose non-common sides form a straight line are supplements.
3. ⊀ *CDB* = ⊀ *CEA*	3. Supplements of equal angles are equal.
4. ⊀ *C* = ⊀ *C*	4. Reflexive law.
5. $\overline{DC}$ ≅ $\overline{EC}$	5. Given.
6. Δ*CDB* ≅ Δ*CEA*	6. ASA ≅ ASA.
7. $\overline{AC}$ ≅ $\overline{BC}$	7. Corresponding parts of congruent triangles are congruent.

 Prove: the median drawn to the base of an isosceles triangle bisects the vertex angle.

 Draw isosceles Δ*ABC* with median CO, as in the figure shown.

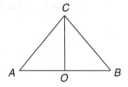

We are asked to prove ⊀ *ACO* ≅ ⊀ *BCO*. By proving Δ*ACO* ≅ Δ*BCO*, we can derive the desired result since the angles are corresponding angles of the two triangles.

Given: isosceles Δ*ABC*; median $\overline{CO}$

Prove: $\overline{CO}$ bisects ⊀ *C*

Statement	Reason
1. $\overline{CO}$ is a median	1. Given.
2. $\overline{AO} \cong \overline{BO}$	2. Definition of median.
3. $\triangle ABC$ is isosceles	3. Given.
4. $\overline{AC} \cong \overline{BC}$	4. Definition of an isosceles triangle.
5. $\overline{CO} \cong \overline{CO}$	5. Reflexive property.
6. $\triangle AOC \cong \triangle BOC$	6. SSS $\cong$ SSS.
7. $\sphericalangle ACO \cong \sphericalangle BCO$	7. Corresponding parts of congruent triangles are congruent.
8. $\overline{CO}$ bisects $\sphericalangle C$	8. A segment that divides an angle into two angles that have equal measure bisects that angle.

Theorem 2

If two angles and a not-included side of one triangle are equal, respectively, to two angles and a not-included side of a second triangle, the triangles are congruent. ($AAS \cong AAS$)

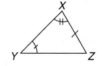

Theorem 3

Corresponding parts of congruent triangles are equal.

Theorem 4

Two right triangles are congruent if the two legs of one right triangle are congruent to the two corresponding legs of the other right triangle.

Theorem 5

If the hypotenuse and an acute angle of one right triangle are equal, respectively, to the hypotenuse and an acute angle of a second right triangle, the triangles are congruent.

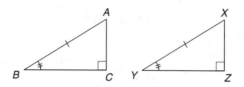

Theorem 6

If the hypotenuse and a leg of one right triangle are equal, respectively, to the hypotenuse and a leg of a second right triangle, the right triangles are congruent.

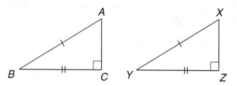

Theorem 7

If a leg and the adjacent acute angle of one right triangle are congruent, respectively, to a leg and the adjacent acute angle of another right triangle, then these two right triangles are congruent.

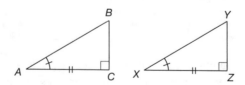

Problem Solving Examples:

Q Prove that the altitudes drawn to the legs of an isosceles triangle are congruent.

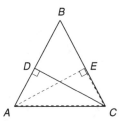

A The accompanying figure shows an isosceles triangle ABC with $\overline{BA} \cong \overline{BC}$, and altitudes $\overline{CD}$ and $\overline{AE}$. By the definition of altitudes, $\overline{CD} \perp \overline{AB}$ and $\overline{AE} \perp \overline{BC}$.

We must prove that $\overline{CD} \cong \overline{AE}$. This can be done by proving $\triangle AEC \cong \triangle CDA$ and employing the corresponding parts rule of congruent triangles.

The congruent triangle postulate that can be best used in this problem is the one which states that two triangles are congruent if two angles and a side opposite one of the angles in one triangle are congruent to the corresponding parts of the other triangle. We shall refer to this rule as the A.A.S. Postulate.

Statement	Reason
1. In $\triangle ABC$, $\overline{BA} \cong \overline{BC}$	1. Given.
2. $\angle BAC \cong \angle BCA$	2. If two sides of a triangle are congruent, the angles opposite these sides are congruent.
3. $\overline{AE} \perp \overline{BC}$, $\overline{CD} \perp \overline{BA}$	3. Given.
4. $\angle CDA$ and $\angle AEC$ are right angles	4. When two perpendicular lines intersect, they form right angles.
5. $\angle CDA \cong \angle AEC$	5. All right angles are congruent.
6. $\overline{AC} \cong \overline{AC}$	6. Reflexive property of congruence.
7. $\triangle ADC \cong \triangle CEA$	7. A.A.S. $\cong$ A.A.S.
8. $\overline{CD} \cong \overline{AE}$	8. Corresponding parts of congruent triangles are congruent.

Q In an isosceles triangle, the sum of the lengths of the perpendiculars drawn to the legs from any point on the base is equal to the length of an altitude drawn to one of the legs. (Hint: Draw $\overline{PF} \perp \overline{AG}$, as shown in the figure.)

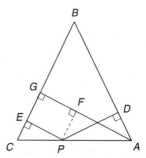

A Let P represent an arbitrary point on the base of $\triangle ABC$. Draw perpendiculars $\overline{PE}$ and $\overline{PD}$, as in the figure, and altitude $\overline{AG}$ to side $\overline{BC}$. We are asked to prove $PE + PD = AG$. We make use of the hint and draw $\overline{PF} \perp \overline{AG}$ to complete $PEGF$.

Since $AG = AF + FG$, we can arrive at the desired conclusion if we prove $PE = FG$ and $PD = AF$. These results can be derived if we prove $PEGF$ is a rectangle and $\triangle APF \cong \triangle PAD$.

Statement	Reason
1. $\overline{PE} \perp \overline{BC}$ and $\overline{AG} \perp \overline{BC}$ (or $\overline{PE} \perp \overline{GE}$ and $\overline{FG} \perp \overline{GE}$)	1. Given.
2. Draw $\overline{PF} \perp \overline{AG}$ or $\overline{PF} \perp \overline{FG}$	2. A perpendicular can be drawn to a line from a given point not on the line.
3. $\overline{PE} \parallel \overline{FG}$	3. Lines perpendicular to the same line are parallel.
4. $\overline{PF} \parallel \overline{GE}$	4. Same as reason 4.
5. $\measuredangle FGE$ is a right angle	5. Perpendicular lines intersect to form a right angle.
6. $PEGF$ is a rectangle	6. A quadrilateral in which all pairs of opposite sides are parallel, which also contains a right angle, is a rectangle.

7. $\overline{PE} \cong \overline{FG}$ or $PE = FG$	7. Opposite sides of a rectangle are congruent.
8. $\triangle ABC$ is isosceles	8. Given.
9. $\angle ECA \cong \angle DAC$	9. Base angles of an isosceles triangle are congruent.
10. $\angle ECA \cong \angle FPA$	10. If parallel lines ($\overline{CG}$ and $\overline{PF}$) are cut by a transversal ($\overline{CA}$), then corresponding angles are congruent.
11. In $\triangle APF$ and $\triangle PAD$ $\angle FPA \cong \angle DAP$	11. Transitive Property of Congruence and steps (9) and (10).
12. $\overline{AP} \cong \overline{PA}$	12. Reflexive Property of Congruence.
13. $\angle PFA$ and $\angle ADP$ are right angles	13. Perpendicular lines intersect in right angles.
14. $\angle PFA \cong \angle ADP$	14. All right angles are congruent.
15. $\triangle APF \cong \triangle PAD$	15. AAS $\cong$ AAS.
16. $\overline{PD} \cong \overline{AF}$ or $PD = AF$	16. Corresponding sides of congruent triangles are congruent.
17. $PE + PD = AF + FG$ or $PE + PD = AG$	17. Substitution Postulate and steps (16) and (7).

4.4   Areas of Triangles

Theorem 1

The area of a triangle is given by the formula $A = \frac{1}{2}bh$, where b is the length of a base and h is the corresponding height of the triangle.

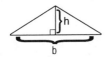

Corollary

If two triangles are congruent, they have the same area.

Theorem 2

The area of a triangle equals one-half the product of any two adjacent sides and the sine of their included angle.

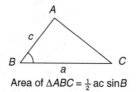

Area of $\triangle ABC = \frac{1}{2} ac \sin B$

Theorem 3

Two triangles with bases of equal length, and altitudes to their bases of equal length, have equal areas.

Theorem 4

Triangles that share the same base and have their vertices on a line parallel to the base, have equal areas.

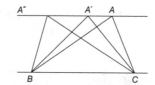

Area of $\triangle A''BC$ = Area of $\triangle A'BC$ = Area of $\triangle ABC$

Theorem 5

In a given triangle, the product of the length of any side and the length of the altitude drawn to that side is equal to the product of the length of any other side and the length of the altitude drawn to that side.

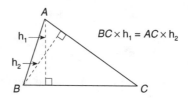

$BC \times h_1 = AC \times h_2$

Theorem 6

The area of a right triangle is equal to one-half the product of the lengths of its two legs.

Theorem 7

A median drawn to a side of a triangle divides the triangle into two triangles that are equal in area.

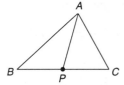

If BP = PC, then
Area of $\triangle$ABP =
Area of $\triangle$APC

Problem Solving Examples:

 In right triangle *ABC*, $\angle C$ measures 90°, $\overline{AB}$ is of length 20 in., and the length of $\overline{AC}$ is 16 in. Find the area of triangle *ABC*.

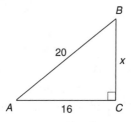

A Assume, as in the accompanying figure, that $\overline{AC}$ is the base of triangle ABC. Since $\angle C$ is a right angle, $\overline{BC}$ is the altitude to the base of the triangle.

The area (*A*) of a right triangle is given by one-half the product of the length of the base (*b*) and the length of the corresponding altitude (*h*). $A = 1/2bh$.

We are not given the length of the altitude, but can calculate it by applying the Pythagorean Theorem to *ABC*. If we let x = the length of the altitude, then,

$$x^2 + (16 \text{ in.})^2 = (20 \text{ in.})^2$$
$$x^2 + 256 \text{ in.}^2 = 400 \text{ in.}^2$$
$$x^2 = 144 \text{ in.}^2$$

which implies

$$x = 12 \text{ in.}$$

The area of *ABC* = $1/2bh$. By substitution,

$$\text{area of } \triangle ABC = 1/2(16 \text{ in.})(12 \text{ in.}) = 96 \text{ in.}^2$$

Therefore, the area of $\triangle ABC$ is 96 sq. in.

 For the figure below, show that $h_1 b_1 = h_2 b_2 = h_3 b_3$.

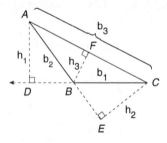

The area of any triangle equals one-half the product of the lengths of its base and the altitude to that base. b_1, b_2, and b_3 designate sides $\overline{BC}$, $\overline{AB}$, and $\overline{AC}$, respectively in $\triangle ABC$. h_1, h_2, and h_3 are altitudes of $\triangle ABC$ drawn from vertices A, C, and B, respectively.

If we consider each side as a base, we obtain:

$$\text{Area of } \triangle ABC = \tfrac{1}{2} h_1 b_1,$$
$$\text{Area of } \triangle ABC = \tfrac{1}{2} h_2 b_2,$$
$$\text{Area of } \triangle ABC = \tfrac{1}{2} h_3 b_3,$$

or

$$\tfrac{1}{2} h_1 b_1 = \tfrac{1}{2} h_2 b_2 = \tfrac{1}{2} h_3 b_3$$

Multiplying by two, we obtain the desired result:

$$h_1 b_1 = h_2 b_2 = h_3 b_3.$$

Theorem 8

The area of a triangle, with sides of lengths a, b and c, is given by the formula

$$A = \sqrt{s(s-a)(s-b)(s-c)}$$

where $s = \tfrac{1}{2}(a+b+c)$, the semiperimeter of the triangle. The above formula is commonly referred to as Heron's formula or Hero's formula.

Theorem 9

The area of an equilaterial triangle is given by the formula

$$A = \frac{x^2 \sqrt{3}}{4}$$

where x is the length of a side of the triangle.

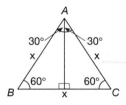

Theorem 10

The areas of two triangles with equal bases have the same ratio as the ratio of their altitudes; and the areas of two triangles with equal altitudes have the same ratio as the ratio of their bases.

Theorem 11

The altitude of an equilateral triangle equals $\frac{\sqrt{3}}{2}$ times the length of a side of the triangle.

Theorem 12

The area of an equilateral triangle equals $\frac{1}{\sqrt{3}}$ times the square of the length of an altitude of the triangle.

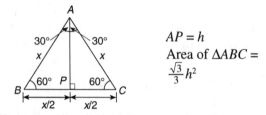

$$AP = h$$
$$\text{Area of } \triangle ABC =$$
$$\frac{\sqrt{3}}{3}h^2$$

Theorem 13

The area of an isosceles triangle with congruent sides length ℓ and an included angle of measure α is: $A = 1/2\,\ell^2 \sin \alpha$. Area is also given by the formula

$$A = h^2 \tan\frac{\alpha}{2} \; ;$$

h is the length of the altitude to the side opposite to the angle α.

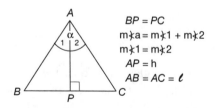

$$BP = PC$$
$$m \angle a = m \angle 1 + m \angle 2$$
$$m \angle 1 = m \angle 2$$
$$AP = h$$
$$AB = AC = \ell$$

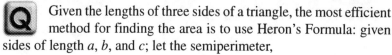

Problem Solving Examples:

Q Given the lengths of three sides of a triangle, the most efficient method for finding the area is to use Heron's Formula: given sides of length a, b, and c; let the semiperimeter,

$$s = \frac{a+b+c}{2}.$$

Then, the area is given by

$$\sqrt{s(s-a)(s-b)(s-c)}.$$

Consider the accompanying figure. $\triangle ABC$ has sides of length a, b, and c. $\overline{CD}$ is the altitude to side $\overline{AB}$, dividing $\overline{AB}$ into segments of length x and $c - x$. Find x.

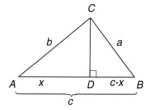

A The length x is determined by the point D. The restriction on D is that $\overline{CD} \perp \overline{AB}$, and thus $\triangle ACD$ and $\triangle CDB$ must be right triangles. Since all right triangles must obey the Pythagorean Theorem, $b^2 = x^2 + CD^2$ and $a^2 = CD^2 + (c - x)^2$. We thus have two simultaneous equations with two unknowns, CD and x. Since we are solving for x, we eliminate $\overline{CD}$. From the first equation $b^2 = x^2 + CD^2$, we obtain $CD^2 = b^2 - x^2$. Substituting this result in the second equation, we obtain:

$$a^2 = (b^2 - x^2) + (c - x)^2 = b^2 - x^2 + c^2 - 2cx + x^2$$

$$a^2 = b^2 + c^2 - 2cx.$$

We isolate x by adding $2cx$ to both sides, subtracting a^2 from both sides, and dividing by $2c$.

$$x = \frac{b^2 + c^2 - a^2}{2c}$$

Find the area of an equilateral triangle whose perimeter is 24.

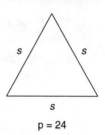

s s

s

p = 24

 The perimeter of an equilateral triangle, whose sides are represented by s, is given by $3s$. In this case, given the perimeter equal to 24, we have $3s = 24$, or $s = 8$.

The area formula for an equilateral triangle is given by the expression

$$A = \frac{s^2 \sqrt{3}}{4}.$$

By substitution,

$$A = \frac{(8)^2 \sqrt{3}}{4} = \frac{64 \sqrt{3}}{4} = 16\sqrt{3}.$$

Therefore, the area of the triangle is $16\sqrt{3}$ sq. units.

CHAPTER 5

Parallelism

5.1 Definitions

Definition 1

A transversal of two or more lines is a line that cuts across these lines in two or more points, one point for each line.

Definition 2

If two lines are cut by a transversal, non-adjacent angles on opposite sides of the transversal, but on the interior of the two lines, are called alternate interior angles.

Definition 3

If two lines are cut by a transversal, non-adjacent angles on opposite sides of the transversal and on the exterior of the two lines are called alternate exterior angles.

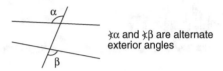

∢α and ∢β are alternate exterior angles

Definition 4

If two lines are cut by a transversal, angles on the same side of the transversal and in corresponding positions with respect to the lines are called corresponding angles.

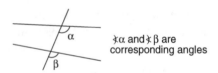

∢α and ∢β are corresponding angles

Definition 5

Two lines are called parallel lines if and only if they are in the same plane (coplanar) and do not intersect. The symbol for parallel, or is parallel to, is ‖: $\overleftrightarrow{AB}$ is parallel to $\overleftrightarrow{CD}$ is written $\overleftrightarrow{AB} \parallel \overleftrightarrow{CD}$.

Definition 6

The distance between two parallel lines is the length of the perpendicular segment from any point on one line to the other line.

$$\ell_1$$
$$\ell_2$$

$\ell_1 \parallel \ell_2$

Problem Solving Examples:

Q Draw two lines and a transversal. Which of the angles in the drawing are corresponding angles? Which are alternate interior angles? Which are alternate exterior angles? Which are consecutive interior angles?

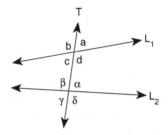

A The situation is shown in the figure. Line T is said to be a transversal of lines L_1 and L_2 provided T, L_1, and L_1 are coplanar and not concurrent.

There are four pairs of corresponding angles: a and α, b and β, c and γ; and d and δ. a and α are a pair of corresponding angles because the position of ∢a with respect to T and L_1 (top left) is the same as the position of ∢α with respect to T and L_2 (top left). Similar logic applies for other pairs of corresponding angles.

There are two pairs of alternate interior angles: β and d; and c and α. The name results from the fact that the pair of angles, say b and δ, are on on "alternate" sides of the transversal and lie in the "interior" of L_1 and L_2.

Angles a and γ and b and δ are alternate exterior angles. The name results from the fact that each pair of alternate exterior angles lies on "alternate" sides of the transversal and exterior to L_1 and L_2.

Angles b and α, and c and δ are consecutive interior angles. The word "consecutive" indicates that the angles lie on the same side of the transversal, and "interior" indicates the location of the angles relative to L_1 and L_2.

Q As shown in the accompanying figure, if $\overline{BD}$ bisects $\sphericalangle ABC$ and $\overline{BC} \cong \overline{CD}$, prove formally that $\overleftrightarrow{CD} \parallel \overleftrightarrow{BA}$.

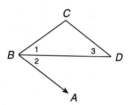

A If two lines are cut by a transversal which forms a pair of alternate interior angles that are congruent, then the two lines are parallel. The transversal of lines $\overleftrightarrow{CD}$ and $\overleftrightarrow{BA}$ is $\overleftrightarrow{BD}$, and $\sphericalangle 2$ and $\sphericalangle 3$ are alternate interior angles of this transversal. Our task is to prove $\sphericalangle 2 \cong \sphericalangle 3$. From the above theorem, it then follows that $\overleftrightarrow{BA} \parallel \overleftrightarrow{CD}$.

Statement	Reason
1. $\overline{BC} \cong \overline{CD}$	1. Given.
2. $\sphericalangle 3 \cong \sphericalangle 1$	2. If two sides of a triangle are congruent, the angles opposite these sides are congruent.
3. $\sphericalangle 1 \cong \sphericalangle 2$	3. An angle bisector divides the angle into two congruent angles.
4. $\sphericalangle 3 \cong \sphericalangle 2$	4. Transitive property of congruence.
5. $\overleftrightarrow{CD} \parallel \overleftrightarrow{BA}$	5. If two lines are cut by a transversal which forms a pair of congruent alternate interior angles, then the two lines are parallel.

5.2   Postulates

Postulate 1

Given a line ℓ and a point P not on line ℓ, there is one and only one line through point P that is parallel to line ℓ.

Postulate 2

Two coplanar lines are either intersecting lines or parallel lines.

Postulate 3

If two (or more) lines are perpendicular to the same line, then they are parallel to each other.

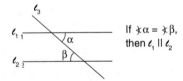

Postulate 4

If two lines are cut by a transversal so that alternate interior angles are equal, the lines are parallel.

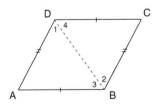

Problem Solving Example:

Prove that if both pairs of opposite sides of a quadrilateral are congruent, then they are also parallel.

Given: Quadrilateral ABCD; $\overline{AB} \cong \overline{CD}$; $\overline{AD} \cong \overline{BC}$

Prove: $\overline{AD} \parallel \overline{BC}$; $\overline{AB} \parallel \overline{CD}$

A In the accompanying figure, the opposite sides of quadrilateral ABCD are congruent. Thus, $\overline{AB} \cong \overline{CD}$ and $\overline{AD} \cong \overline{BC}$. We must show $\overline{CD} \parallel \overline{AB}$ and $\overline{AD} \parallel \overline{BC}$.

To do this, we draw diagonal $\overline{DB}$. Remember that if alternate interior angles are congruent, the two lines are parallel. Thus, to show $\overline{AB} \parallel \overline{CD}$, we prove $\sphericalangle 3 \cong \sphericalangle 4$. To show $\sphericalangle 3 \cong \sphericalangle 4$, we prove $\triangle ADB \cong \triangle CBD$ by the SSS Postulate. Thus, by corresponding parts, $\sphericalangle 3 \cong \sphericalangle 4$ and $\overline{AB} \parallel \overline{BC}$.

By corresponding parts, we can also say $\sphericalangle 1 \cong \sphericalangle 2$. Since $\sphericalangle 1$ and $\sphericalangle 2$ are alternate interior angles of $\overleftrightarrow{AD}$ and $\overleftrightarrow{BC}$, it follows that $\overline{AD} \parallel \overline{BC}$.

Statement	Reason
1. Quadrilateral ABCD; $\overline{AB} \cong \overline{CD}$; $\overline{AD} \cong \overline{CD}$	1. Given.
2. $\overline{DB} \cong \overline{DB}$	2. A segment is congruent to itself.
3. $\triangle ADB \cong \triangle CBD$	3. The SSS Postulate.
4. $\sphericalangle 1 \cong \sphericalangle 2$ $\sphericalangle 3 \cong \sphericalangle 4$	4. Corresponding angles of congruent triangles are congruent.
5. $\overline{AD} \parallel \overline{BC}$, $\overline{AB} \parallel \overline{CD}$	5. If two coplanar lines are cut by a transversal such that the alternate interior angles are congruent, then the lines are parallel.

5.3 Theorems

Theorem 1

If two lines are parallel to the same line, then they are parallel to each other.

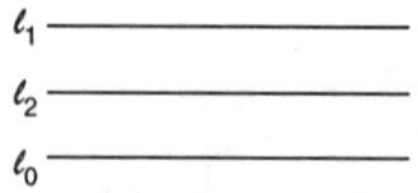

ℓ_1 ———————————

ℓ_2 ———————————

ℓ_0 ———————————

If $\ell_1 \parallel \ell_0$ and $\ell_2 \parallel \ell_0$, then $\ell_1 \parallel \ell_2$

Theorem 2

If a line is perpendicular to one of two parallel lines, then it is perpendicular to the other line, too.

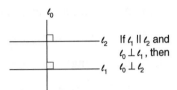

If $\ell_1 \parallel \ell_2$ and
$\ell_0 \perp \ell_1$, then
$\ell_0 \perp \ell_2$

Theorem 3

If two lines being cut by a transversal form congruent corresponding angles, then the two lines are parallel.

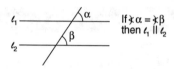

If $\sphericalangle \alpha = \sphericalangle \beta$
then $\ell_1 \parallel \ell_2$

Theorem 4

If two lines being cut by a transversal form interior angles on the same side of the transversal that are supplementary, then the two lines are parallel.

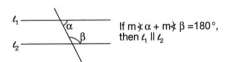

If $m \sphericalangle \alpha + m \sphericalangle \beta = 180°$,
then $\ell_1 \parallel \ell_2$

Theorem 5

If a line is parallel to one of two parallel lines, it is also parallel to the other line.

If $\ell_1 \parallel \ell_2$
and $\ell_0 \parallel \ell_1$, then
$\ell_0 \parallel \ell_2$

Theorem 6

If two parallel lines are cut by a transversal, then:

(A) The alternate interior angles are congruent.

(B) The corresponding angles are congruent.

(C) The consecutive interior angles are supplementary.

(D) The alternate exterior angles are congruent.

Theorem 7

Parallel lines are always the same distance apart.

Problem Solving Examples:

 Prove that if L_1, L_2, and L_3 are lines in a plane, M, such that $L_1 \perp L_3$ and $L_2 \perp L_3$, then $L_1 \parallel L_2$ or $L_1 = L_2$. (see figure)

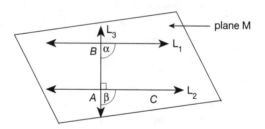

We will use the fact that if the corresponding angles of two lines cut by a transversal are congruent, then the two lines are parallel. From this, we can show that $L_1 \parallel L_2$ (or $L_1 = L_2$).

Let point A be the intersection of L_1 and L_3 and point B be the intersection of L_3 and L_2. Two cases must be considered:

(1) A not coincident with B ($A \neq B$)

(2) A coincides with B ($A = B$)

(1) We first assume $A \neq B$.

Note that angles α and β are corresponding angles for lines $L_1 \parallel L_2$ cut by transversal, L_3

From the statement of the problem, $L_1 \perp L_3$ and $L_2 \perp L_3$ and thus both α and β are right angles. Hence, $\sphericalangle \alpha \cong \sphericalangle \beta$. This implies that $L_1 \parallel L_2$.

(2) However, suppose points A and B coincide in the figure so that A equals B. Then, if the given facts still hold, both L_1 and L_2 must be perpendicular to L_3 at the same point. Since there is only one line perpendicular to a given line at a given point, L_1 and L_2 must coincide (i.e., $L_1 = L_2$).

 If line $\overleftrightarrow{AB}$ is parallel to line $\overleftrightarrow{CD}$ and line $\overleftrightarrow{EF}$ is parallel to line $\overleftrightarrow{GH}$, prove that $m\sphericalangle 1 = m\sphericalangle 2$.

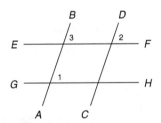

 To show $\sphericalangle 1 \cong \sphericalangle 2$, we relate both to $\sphericalangle 3$. Because $\overline{EF} \parallel \overline{GH}$, corresponding angles 1 and 3 are congruent. Since $\overline{AB} \parallel \overline{CD}$, corresponding angles 3 and 2 are congruent. Because both $\sphericalangle 1$ and $\sphericalangle 2$ are congruent to the same angle, it follows that $\sphericalangle 1 \cong \sphericalangle 2$.

Statement	Reason
1. $\overleftrightarrow{EF} \parallel \overleftrightarrow{GH}$	1. Given.
2. $m\sphericalangle 1 = m\sphericalangle 3$	2. If two parallel lines are cut by a transversal, corresponding angles are of equal measure.
3. $\overleftrightarrow{AB} \parallel \overleftrightarrow{CD}$	3. Given.
4. $m\sphericalangle 2 = m\sphericalangle 3$	4. If two parallel lines are cut by a transversal, corresponding angles are equal in measure.
5. $m\sphericalangle 1 = m\sphericalangle 2$	5. If two quantities are equal to the same quantity, they are equal to each other.

5.4   Corollaries

Corollary 1

If a line intersects one of two parallel lines, it intersects the other line also.

Corollary 2

If two lines are cut by a transversal so that alternate interior angles are not equal, the lines are not parallel.

Corollary 3

If two lines are cut by a transversal so that corresponding angles are not equal, the lines are not parallel.

Corollary 4

If two lines are cut by a transversal so that two interior angles on the same side of the transversal are not supplementary, the lines are not parallel.

Corollary 5

If two nonparallel lines are cut by a transversal, the pairs of alternate interior angles are not equal.

Corollary 6

If line A is perpendicular to one of two parallel lines, and if another line B is perpendicular to the second of the two parallel lines, then lines A and B are parallel to each other.

Corollary 7

If three or more parallel lines intercept congruent segments on one transversal, then they intercept congruent segments on any transversal.

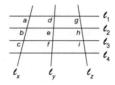

For example:

Given: Four parallel lines, ℓ_1, ℓ_2, ℓ_3 and ℓ_4 intersected by three transversals ℓ_x, ℓ_y and ℓ_z, with the lengths of the transversals between parallel lines represented by the letters a through i as shown.

If $a = b = c$, then $d = e = f$ and $g = h = i$.

Problem Solving Examples:

   Given $\overline{AB} \parallel \overline{DC}$ and $\overline{AB} \cong \overline{CD}$.

Prove: $\angle A \cong \angle C$.

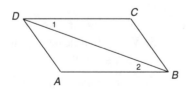

   $\angle A$ and $\angle C$ are corresponding angles of $\triangle BDC$ and $\triangle DBA$. Thus, to prove $\angle A \cong \angle C$, we first prove $\triangle BDC \cong \triangle DBA$ by the SAS Postulate.

Statement	Reason
1. $\overline{AB} \parallel \overline{DC}$	1. Given.
2. $\angle 1 \cong \angle 2$	2. If two parallel lines are cut by a transversal, alternate interior angles are congruent.
3. $\overline{AB} \cong \overline{CD}$	3. Given.
4. $\overline{DB} \cong \overline{DB}$	4. Reflexive property.
5. $\triangle ADB \cong \triangle CBD$	5. SAS $\cong$ SAS.
6. $\angle A \cong \angle C$	6. Corresponding parts of congruent triangles are congruent.

 Given: $\overline{AC}$ and $\overline{EB}$ bisect each other at D.
Prove: $\overline{AE} \parallel \overline{BC}$.

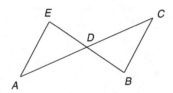

 To show two lines are parallel, it is sufficient to show that a pair of alternate interior angles, such as $\angle A$ and $\angle C$, are congruent.

We first prove $\triangle AED \cong \triangle CBD$ by the SAS postulate. Because corresponding parts of congruent triangles are congruent, $\angle A \cong \angle C$, and thus the lines are parallel.

Statement	Reason
1. $\overline{AC}$ and $\overline{EB}$ bisect each other at D	1. Given.
2. $\overline{ED} \cong \overline{BD}$; $\overline{AD} \cong \overline{CD}$	2. Definition of bisector.
3. $\angle EDA \cong \angle BDC$	3. Vertical angles are congruent.
4. $\triangle EDA \cong \triangle BDC$	4. SAS $\cong$ SAS.
5. $\angle A \cong \angle C$	5. Corresponding parts of congruent triangles are congruent.
6. $\overline{AE} \parallel \overline{BC}$	6. If two lines are cut by a transversal so that alternate interior angles are congruent, the lines are parallel.

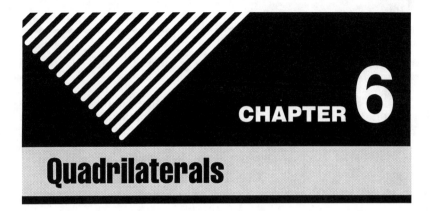

CHAPTER 6

Quadrilaterals

6.1 Parallelograms

Definition 1

A quadrilateral is a polygon with four sides.

Definition 2

A parallelogram is a quadrilateral whose opposite sides are parallel.

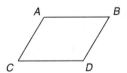

Definition 3

Two angles that have their vertices at the endpoints of the same side of a parallelogram are called consecutive angles.

Definition 4

The perpendicular segment connecting any point of a line containing one side of the parallelogram to the line containing the opposite side of the parallelogram is called an altitude of the parallelogram.

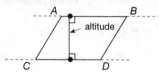

Definition 5

A diagonal of a polygon is a line segment joining any two non-consecutive vertices.

Problem Solving Examples:

 If $ABCD$ is a quadrilateral such that $\overline{AB} \cong \overline{CD}$ and $\overline{AD} \cong \overline{BC}$, then prove that $ABCD$ is a parallelogram. (see figure)

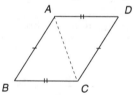

 Our strategy is to show that $\triangle ADC \cong \triangle CBA$ by the SSS Postulate. This will tell us that $\sphericalangle BAC \cong \sphericalangle DCA$. Since these are alternate interior angles for segments $\overline{AB}$ and $\overline{CD}$, we will have shown that $\overline{AB} \parallel \overline{CD}$. A similar procedure will yield the fact that $\overline{AD} \parallel \overline{BC}$. By definition, this will mean that $ABCD$ is a parallelogram.

By the given facts, $\overline{AB} \cong \overline{CD}$ and $\overline{AD} \cong \overline{BC}$; $\overline{AC}$ is shared by $\triangle ADC$ and $\triangle CBA$. By reflexivity, $\overline{AC} \cong \overline{AC}$. By the SSS Postulate, then, $\triangle ADC \cong \triangle CBA$. By corresponding parts,

$$\angle BAC \cong \angle DCA \tag{1}$$

and

$$\angle BCA \cong \angle DAC. \tag{2}$$

But the set of angles in (1) are alternate interior angles with respect to segments $\overline{BA}$ and $\overline{DC}$. Hence, $\overline{BA} \parallel \overline{DC}$. Similarly, the set of angles in (2) are alternate interior angles with respect to segments $\overline{AD}$ and $\overline{BC}$. Since the angles are congruent, $\overline{AD} \parallel \overline{BC}$.

$ABCD$ has both pairs of opposite sides congruent and parallel and is, by definition, a parallelogram.

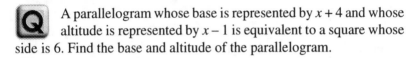 A parallelogram whose base is represented by $x + 4$ and whose altitude is represented by $x - 1$ is equivalent to a square whose side is 6. Find the base and altitude of the parallelogram.

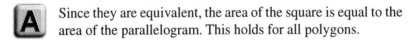 Since they are equivalent, the area of the square is equal to the area of the parallelogram. This holds for all polygons.

The area of the square equals the square of the length of any side s. Therefore, $A_s = s^2$. By substitution, $s = 6$, and $A_s = 6^2 = 36$.

The area of a parallelogram, A_p, is equal to the product of lengths of any base and its corresponding altitude. In this example, $A_p = (x + 4)(x - 1) = x^2 + 3x - 4$.

The above rule for equivalent polygons allows us to set $A_p = A_s$ and solve for x. Hence,

$$x^2 + 3x - 4 = 36$$
$$x^2 + 3x - 40 = 0$$

By factoring,

$$(x + 8)(x - 5) = 0.$$

Then, either $x + 8 = 0$ or $x - 5 = 0$, which implies $x = -8$ or $x = 5$. The negative answer is rejected because it has no geometric significance. Therefore,   Base $= x + 4 = 5 + 4 = 9$.

Altitude $= x - 1 = 5 - 1 = 4$.

Therefore, base of parallelogram $= 9$, altitude $= 4$.

Theorem 1

A diagonal of a parallelogram divides the parallelogram into two congruent triangles.

Theorem 2

Consecutive angles of a parallelogram are supplementary.

Theorem 3

If both pairs of opposite sides of a quadrilateral are equal, then the quadrilateral is a parallelogram.

Theorem 4

The diagonals of a parallelogram bisect each other.

Theorem 5

If two opposite sides of a quadrilateral are both parallel and equal, the quadrilateral is a parallelogram.

$\overline{BC} \parallel \overline{AD}$

Problem Solving Examples:

In the figure, $ABCD$ is a parallelogram with diagonals $\overline{AC}$ and $\overline{BD}$. $\measuredangle ABC$ is an obtuse angle. Prove that $\overline{AC} > \overline{BD}$.

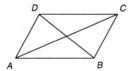

A   The main theorem to be applied here is the one which states that if two sides of one triangle are congruent, respectively, to two sides of another triangle, and the included angles are not congruent, then the triangle which has the included angle of larger measure has the third side of greater length.

$\overline{AC}$ and $\overline{BD}$ are in different triangles. We will show $\overline{AC} > \overline{BD}$, under the above theorem, by showing two sides in $\triangle ABC$, $\overline{BC}$ and $\overline{AB}$, are congruent to two sides in $\triangle BAD$, $\overline{AD}$ and $\overline{BA}$ and that the measure of the included angle ABC is greater than the measure of the included angle *BAD*.

Statement	Reason
1. *ABDC* is a parallelogram	1. Given.
2. $\overline{AD} \cong \overline{BC}$	2. Opposite sides of a parallelogram are congruent.
3. $\angle BAD$ is supplementary to $\angle ABC$	3. Consecutive angles of a parallelogram are supplementary.
4. $\angle ABC$ is an obtuse angle	4. Given.
5. $\angle BAD$ is an acute angle	5. The supplement of an obtuse angle is an acute angle. (see step 3)
6. $m\angle ABC > m\angle BAD$	6. By definition, the measure of an obtuse angle is greater than the measure of an acute angle.
7. $\overline{AB} \cong \overline{BA}$	7. Reflexive Property of Congruence.
8. $\overline{AC} > \overline{BD}$	8. In two triangles, if two sides of one triangle are congruent, respectively, to two sides of the other, and the included angles are not congruent, then the triangle which has the included angle of larger measure has the greater third side.

Prove that a quadrilateral, in which one pair of opposite sides are both congruent and parallel, is a parallelogram.

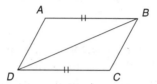

A In quadrilateral *ABCD* (as shown in the figure) assume $\overline{AB} \cong \overline{CD}$ and $\overline{AB} \parallel \overline{CD}$. We shall prove that $\triangle DAB \cong \triangle BCD$ by the SAS Postulate. We will then know that $\overline{DA} \cong \overline{BC}$. Since we already know that $\overline{AB} \cong \overline{CD}$, we will have shown that both pairs of opposite sides of the quadrilateral are congruent, implying that *ABCD* is a parallelogram.

As given, $\overline{AB} \cong \overline{CD}$. Furthermore, since $\overline{AB} \parallel \overline{CD}$, then $\sphericalangle ABD \cong \sphericalangle CDB$. This follows because these angles are alternate interior angles of parallel segments cut by a transversal. Lastly, $\overline{DB}$ is shared by both triangles, therefore $\overline{DB} \cong \overline{DB}$. By the SAS Postulate, $\triangle DAB \cong \triangle BCD$. This implies that $\overline{DA} \cong \overline{BC}$.

We have shown that both pairs of opposite sides of quadrilateral *ABCD* are congruent. This implies that *ABCD* is a parallelogram.

Theorem 6

If the diagonals of a quadrilateral bisect each other, then the quadrilateral is a parallelogram.

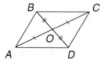

Theorem 7

Opposite sides of a parallelogram are equal.

Theorem 8

Non-consecutive angles of a parallelogram are equal.

Theorem 9

If both pairs of opposite angles of a quadrilateral are congruent, the quadrilateral is a parallelogram.

Theorem 10

If one angle of a quadrilateral is congruent to the opposite angle and one side is parallel to the opposite side, then the quadrilateral is a parallelogram.

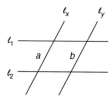

$$\overline{AB} \parallel \overline{CD}$$

Theorem 11

If one angle of a quadrilateral is congruent to its opposite, and one side is congruent to its opposite, then the quadrilateral is a parallelogram.

Corollary

Segments of parallel lines intercepted between parallel lines are congruent.

If $\ell_1 \parallel \ell_2$ and $\ell_x \parallel \ell_y$, then $a = b$.

Problem Solving Example:

 Let *ABCD* be a parallelogram. Prove that △*ABD* ≅ △*CDB*. (see figure)

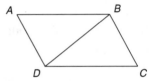

 We shall use the definition of a parallelogram, and the properties of parallel lines, to show that △*ABD* ≅ △*CDB* by the ASA Postulate.

In a parallelogram, each pair of opposite sides is parallel (i.e., $\overline{AB}$ ‖ $\overline{DC}$ and $\overline{AD}$ ‖ $\overline{BC}$). Because $\overline{AB}$ ‖ $\overline{DC}$, ∡*ABD* ≅ ∡*CDB* since they are alternate interior angles formed by transversal $\overline{BD}$. Similarly, $\overline{AD}$ ‖ $\overline{BC}$, which implies that alternate interior angles *ADB* and *CBD* are congruent. $\overline{DB}$ is common to both △*ABD* and △*CDB* and, by reflexivity $\overline{DB}$ ≅ $\overline{DB}$. Hence, by the ASA Postulate, △*ABD* ≅ △*CDB*.

6.2 Rectangles

Definition

A rectangle is a parallelogram with one right angle.

Theorem 1

All angles of a rectangle are right angles.

Theorem 2

The diagonals of a rectangle are equal.

Theorem 3

If the diagonals of a parallelogram are equal, the parallelogram is a rectangle.

Theorem 4

If a quadrilateral has four right angles, then it is a rectangle.

Theorem 5

If a parallelogram is inscribed within a circle, then it is a rectangle.

In a circle O with inscribed quadrilateral $ABCD$, quadrilateral $ABCD$ is a parallelogram; therefore, quadrilateral $ABCD$ is a rectangle.

Problem Solving Examples:

 Prove that a rectangle is a parallelogram.

 A rectangle is a quadrilateral with four right angles. We will show that $ABCD$ is a parallelogram by analyzing the pairs of angles ABC and BCD, and DAB and ABC.

Since $\angle ABC$, $\angle BCD$, and $\angle DAB$ are all right angles,

$$m \sphericalangle ABC = 90° \tag{1}$$
$$m \sphericalangle BDC = 90° \tag{2}$$
$$m \sphericalangle DAB = 90° \tag{3}$$

Adding (1) and (2) shows that $\sphericalangle ABC$ and $\sphericalangle BCD$ are supplementary:

$$m \sphericalangle ABC + m \sphericalangle BCD = 180° \tag{4}$$

Adding (1) and (3) shows that $\sphericalangle ABC$ and $\sphericalangle DAB$ are supplementary:

$$m \sphericalangle ABC + m \sphericalangle DAB = 180°. \tag{5}$$

Now, if the consecutive interior angles of two lines crossed by a transversal sum to 180°, then the two lines are parallel.

$\sphericalangle ABC$ and $\sphericalangle BCD$ are consecutive interior angles of line segments $\overline{AB}$ and $\overline{DC}$. Also, $\sphericalangle ABC$ and $\sphericalangle DBA$ are consecutive interior angles of line segments $\overline{AD}$ and $\overline{BC}$. Using these facts, plus (4) and (5), we conclude that $\overline{AD} \parallel \overline{BC}$ and $\overline{AB} \parallel \overline{DC}$. By definition this means that $ABCD$, a rectangle, is a parallelogram.

Q A rectangle is inscribed in a circle whose radius is 5 inches. The base of the rectangle is 8 inches. Find the area of the rectangle. (see figure)

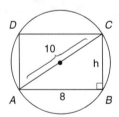

A The diagonal of the rectangle is a diameter of the circle and as such, each of the two triangles shown in the figure is inscribed in a semi-circle. Therefore, they are right triangles.

Since $\triangle BAC$ is a right triangle, we can use the Pythagorean Theorem to determine the altitude, h, of $ABCD$, as indicated in the figure.

We will need this to calculate the area of the rectangle. Therefore,

$$h^2 + (8 \text{ in.})^2 = (10 \text{ in.})^2$$
$$h^2 + 64 \text{ in.}^2 = 100 \text{ in.}^2$$
$$h^2 = 36 \text{ in.}^2 \quad \text{and}$$
$$h = 6 \text{ in.}$$

Now, Area = bh. Let $b = 8$ in., $h = 6$ in. By substitution,

$$A = (8 \times 6)\text{in.}^2 = 48 \text{ in.}^2$$

Therefore, the area of the inscribed rectangle is 48 sq. in.

6.3   Rhombi

Definition

A rhombus is a parallelogram with two adjacent sides equal.

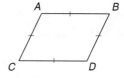

Theorem 1

All sides of a rhombus are equal.

Theorem 2

The diagonals of a rhombus are perpendicular to each other.

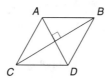

Theorem 3

The diagonals of a rhombus bisect the angles of the rhombus.

Theorem 4

If the diagonals of a parallelogram are perpendicular, the parallelogram is a rhombus.

Theorem 5

If a quadrilateral has four equal sides, then it is a rhombus.

Theorem 6

A parallelogram is a rhombus if either diagonal of the parallelogram bisects the angles of the vertices it joins.

Problem Solving Examples:

 In the figure shown, $\overrightarrow{BF}$ bisects angle CBA. $\overline{DE} \parallel \overrightarrow{BA}$ and $\overline{GE}$ $\parallel \overrightarrow{BC}$. Prove *GEDB* is a rhombus.

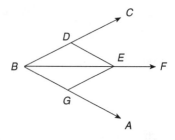

 We want to show that *GEDB* is a parallelogram which has congruent adjacent sides, since this is, by definition, a rhombus.

We will prove that *GEDB* is a parallelogram, and then prove $\triangle BEG$ $\cong \triangle BED$ in order to deduce that adjacent sides $\overline{GE}$ and $\overline{DE}$ are congruent.

Statement	Reason
1. $\overline{DE} \parallel \overrightarrow{BA}$, $\overline{GE} \parallel \overrightarrow{BC}$	1. Given.
2. *GEDB* is a parallelogram	2. A quadrilateral in which all opposite sides are parallel is a parallelogram.
3. $\overrightarrow{BF}$ is the angle bisector of ∢*CBA*	3. Given.
4. ∢*DBE* ≅ ∢*GBE*	4. An angle bisector divides the angle into two congruent angles.
5. $\overline{BG} \cong \overline{DE}$	5. Opposite sides of a parallelogram are congruent.
6. $\overline{BE} \cong \overline{BE}$	6. Reflexive Property of Congruence.
7. Δ*BEG* ≅ Δ*BED*	7. S.A.S. ≅ S.A.S.
8. $\overline{GE} \cong \overline{DE}$	8. Corresponding sides of congruent triangles are congruent.
9. *GEDB* is a rhombus	9. A rhombus is a parallelogram with adjacent sides congruent.

 Let *ABCD* be a rhombus. Prove that diagonal $\overline{AC}$ bisects ∢*A*.

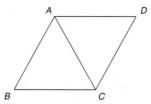

The figure shows rhombus *ABCD* with diagonal $\overline{AC}$. To show that ∢*A* is bisected by $\overline{AC}$, we prove that ∢*BAC* ≅ ∢*DAC*. This can be accomplished by showing that Δ*BAC* ≅ Δ*DAC*.

A rhombus is a quadrilateral, all of whose sides are congruent. Hence, $\overline{BA} \cong \overline{DA}$ and $\overline{BC} \cong \overline{DC}$. $\overline{AC} \cong \overline{AC}$, by the reflexive property. Therefore, by the SSS Postulate, Δ*BAC* ≅ Δ*DAC*, which implies that ∢*BAC* ≅ ∢*DAC*. Hence, $\overline{AC}$ bisects ∢*A*.

6.4   Squares

Definition

A square is a rhombus with a right angle.

Theorem 1

A square is an equilateral quadrilateral.

Theorem 2

A square has all the properties of parallelograms and rectangles.

Theorem 3

A rhombus is a square if one of its interior angles is a right angle.

Theorem 4

In a square, the measure of either diagonal can be calculated by multiplying the length of any side by the square root of 2.

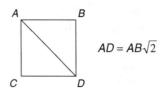

$$AD = AB\sqrt{2}$$

Problem Solving Examples:

Q In the accompanying figure, $\triangle ABC$ is given to be an isosceles right triangle with $\angle ABC$ a right angle and $\overline{AB} \cong \overline{BC}$. Line segment $\overline{BD}$, which bisects $\overline{CA}$, is extended to E, so that $\overline{BD} \cong \overline{DE}$. Prove $BAEC$ is a square.

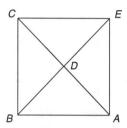

A A square is a rectangle in which two consecutive sides are congruent. This definition will provide the framework for the proof in this problem. We will prove that $BAEC$ is a parallelogram that is specifically a rectangle with consecutive sides congruent, namely a square.

Statement	Reason
1. $\overline{BD} \cong \overline{DE}$ and $\overline{AD} \cong \overline{DC}$	1. Given (BD bisect CA).
2. *BAEC* is a parallelogram	2. If diagonals of a quadrilateral bisect each other, then the quadrilateral is a parallelogram.
3. $\angle ABC$ is a right angle	3. Given.
4. *BAEC* is a rectangle	4. A parallelogram, one of whose angles is a right angle, is a rectangle.
5. $\overline{AB} \cong \overline{BC}$	5. Given.
6. *BAEC* is a square	6. If a rectangle has two congruent consecutive sides, then the rectangle is a square.

Given: Square *PQRS*, with *N* on $\overline{RS}$ so that $\overline{TS} \cong \overline{SN}$.
Prove: $m \sphericalangle STN = 3(m \sphericalangle NTR)$.

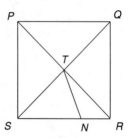

 Here, we can solve for the exact values of $\sphericalangle STN$ and $\sphericalangle NTR$ and then compare. We are given that ΔSTN is isosceles. Because *PQRS* is a rhombus (every square is a rhombus) diagonal SQ bisects right angle $\sphericalangle PSR$ and $m \sphericalangle TSR = \frac{1}{2}(90°) = 45°$. $\sphericalangle STN$ is a base angle of isosceles ΔSTN, and thus $\sphericalangle STN \cong \sphericalangle SNT$. Since $180° = m \sphericalangle TSN + m \sphericalangle STN + m \sphericalangle SNT$ and $m \sphericalangle SNT = m \sphericalangle SNT$, by substitution, $2m \sphericalangle STN = 180° - 45°$, and thus $m \sphericalangle STN = \frac{135°}{2} = 67.5°$.

Since the diagonals of a square are perpendicular to each other, $\sphericalangle STN$ and $\sphericalangle NTR$ are complementary, and $m \sphericalangle NTR = 90° - 67.5° = 22.5°$. $67.5° = 3(22.5°)$, so $m \sphericalangle STN = 3(m \sphericalangle NTR)$.

Statement	Reason
1. Square *PQRS*, with *N* in RS, so that $\overline{TS} \cong \overline{SN}$	1. Given.
2. $\overline{QS}$ bisects $\sphericalangle PSR$	2. The diagonals of a rhombus bisect the angles.
3. $m \sphericalangle TSN = \frac{1}{2} m \sphericalangle PSR$	3. The bisector of an angle divides an angle into two equal parts.
4. $m \sphericalangle PSR = 90°$	4. The angles of a square are right angles.
5. $m \sphericalangle TSN = \frac{1}{2} 90° = 45°$	5. Substitution Postulate.
6. $m \sphericalangle TSN + m \sphericalangle STN + m \sphericalangle SNT = 180°$	6. The angle sum of a triangle is 180°.

7. $m\sphericalangle STN = m\sphericalangle SNT$	7. The base angles of an isosceles triangle are congruent.
8. $2m\sphericalangle STN + 45° = 180°$	8. Substitution Postulate.
9. $m\sphericalangle STN = \dfrac{180° - 45°}{2}$ $= 67.5°$	9. Subtraction and Division Properties of Equality.
10. $\overline{QS} \perp \overline{PR}$	10. The diagonals of a rhombus are perpendicular to each other.
11. $m\sphericalangle STN + m\sphericalangle NTR = 90°$	11. Two angles that form a right angle are complementary.
12. $m\sphericalangle NTR = 90° - 67.5°$ $= 22.5°$	12. Substitution Postulate.
13. $67.5° = 3 \cdot 22.5°$	13. Factoring Postulate.
14. $m\sphericalangle STN = 3(m\sphericalangle NTR)$	14. Substitution Postulate.

6.5 Trapezoids

Definition 1

A trapezoid is a quadrilateral with two and only two sides parallel. The parallel sides of a trapezoid are called bases.

Definition 2

The median of a trapezoid is the line joining the midpoints of the non-parallel sides.

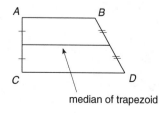

median of trapezoid

Definition 3

The perpendicular segment connecting any point in the line containing one base of the trapezoid to the line containing the other base is the altitude of the trapezoid.

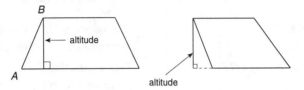

Definition 4

An isosceles trapezoid is a trapezoid whose non-parallel sides are equal. A pair of angles including only one of the parallel sides is called a pair of base angles.

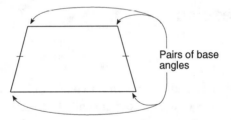

Pairs of base angles

Problem Solving Example:

The lengths of the bases of an isosceles trapezoid are 8 and 14, and each of the base angles measures 45°. Find the length of the altitude of the trapezoid.

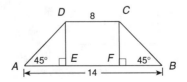

A　　As can be seen in the figure, it is helpful to draw both altitudes $\overline{DE}$ and $\overline{CF}$. *DCFE* is a rectangle because $\overline{DC} \parallel \overline{AB}$ (given by the definition of a trapezoid), $\overline{DE} \parallel \overline{CD}$ (they are both $\perp$ to the same line) and $\angle E$ is a right angle ($\overline{DE} \perp \overline{AB}$). Opposite sides of a rectangle are congruent. Therefore, $\overline{DC} = \overline{EF} = 8$.

$\triangle AED \cong \triangle BFC$ because $\angle DEA \cong \angle CFB$ (both are right angles).

$\angle DAE \cong \angle CBF$, and $\overline{DA} \cong \overline{BC}$. (The last two facts come from the definition of an isosceles trapezoid). Therefore, by corresponding parts, $\overline{AE} \cong \overline{BF}$. Then, $\overline{AE} = \frac{1}{2}(\overline{AB} - \overline{EF})$ and, substituting $\overline{AE} = \frac{1}{2}(14 - 8)$ $= \frac{1}{2}(6) = 3$.

In $\triangle AED$, $m\angle E = 90$ and $m\angle A = 45$. By the angle sum postulate for triangles, $m\angle D = 45$. Since $\triangle AED$ has two angles of equal measure, it must be an isosceles triangle. Therefore, $\overline{AE} = \overline{DE} = 3$.

Therefore, the length of the altitude of trapezoid *ABCD* is 3.

Theorem 1

The median of a trapezoid is parallel to the bases and equal to one-half their sum.

Theorem 2

The base angles of an isosceles trapezoid are equal.

Theorem 3

The diagonals of an isosceles trapezoid are equal.

Theorem 4

The opposite angles of an isosceles trapezoid are supplementary.

Theorem 5

If one pair of base angles of a trapezoid are congruent, then the trapezoid is isosceles.

Theorem 6

A trapezoid is isosceles if any angle and its opposite are supplementary.

Theorem 7

If the diagonals of a trapezoid are congruent then the trapezoid is isosceles.

Problem Solving Example:

 Prove that the line containing the median of a trapezoid bisects any altitude of the trapezoid.

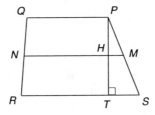

 The median of a trapezoid is the line segment joining the midpoints of the nonparallel sides. Since the median is everywhere halfway between the bases, it is everywhere equidistant from the two bases and, as such, is parallel to them. The nonparallel sides and any altitude of a trapezoid are all transversals of the parallel bases and median.

There is a theorem that states if three or more parallel lines intercept congruent segments on one transversal, then they intercept congruent segments on any other transversal. Since parallel lines $\overline{QP}$, $\overline{NM}$, and $\overline{RS}$ cut equal segments on $\overline{QR}$ and $\overline{PS}$, they cut $\overline{PT}$ equally.

Given: Trapezoid PQRS with altitude $\overline{PT}$ intersecting median $\overline{MN}$ at *H*.

Prove: $\overline{MN}$ bisects $\overline{PT}$.

Statement	Reason
1. Trapezoid *PQRS* with altitude $\overline{PT}$ meeting median $\overline{MN}$ at *H*	1. Given.
2. $\overline{MN} \parallel \overline{SR}$	2. The median of a trapezoid is parallel to the bases.
3. *M* is the midpoint of $\overline{PS}$	3. Definition of the median of a triangle.
4. $\overline{PM} \cong \overline{MS}$	4. Definition of the midpoint of a line segment.
5. $\overline{PH} \cong \overline{HT}$	5. If three or more parallel lines intercept congruent segments on a transversal, then they intercept congruent segments on any other transversal.
6. $\overline{MN}$ bisects $\overline{PT}$	6. Definition of bisection.

CHAPTER 7

Geometric

Inequalities

7.1 Postulates and Theorems

Postulate 1

A quantity may be substituted for its equal in any inequality.

Postulate 2

A whole quantity is greater than any of its parts.

Postulate 3

The relation "<" is transitive; that is, if $a < b$ and $b < c$, then $a < c$.

Postulate 4

If the same quantity is added to both sides of an inequality, the sums are still unequal and in the same order.

(If $a < b$, then $a + c < b + c$.)

Postulate 5

If equal quantities are added to unequal quantities, the sums are still unequal and in the same order.

(If $a < b$ and $c = d$, then $a + c < b + d$.)

Postulate 6

If unequal quantities are added to unequal quantities of the same order, the sums are unequal quantities and in the same order.

(If $a < b$ and $c < d$, then $a + c < b + d$.)

Problem Solving Examples:

Q In triangle ABC, angles ABC and ACB are divided by $\overline{BD}$ and $\overline{CE}$, respectively. This results in $m\angle BCE > m\angle DBC$ and $m\angle ACE > m\angle ABD$. Prove: $m\angle ACB > m\angle ABC$.

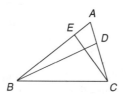

A The Addition Postulate of Inequalities states that when unequal quantities are added to unequal quantities of the same order, then the sums are unequal in the same order. (The three main orders are greater than/less than/equal to.) In the given problem, angles *ABC* and *ACB* are each divided into two parts. Each part of $\angle ACB$ is greater than a part of $\angle ABC$. By the above-mentioned Postulate, $m\angle ACB > m\angle ABC$.

Given: $\triangle ABC$; $m\angle BCE > m\angle DBC$; $m\angle ACE > m\angle ABD$.

Prove: $m\angle ACB > m\angle ABC$.

Statement	Reason
1. $m \angle BCE > m \angle DBC$ $m \angle ACE > m \angle ABD$	1. Given.
2. $m \angle BCE + m \angle ACE$ $m \angle DBC + m \angle ABD$	2. Addition Postulate of Inequality.
3. $m \angle ACB = m \angle BCE + m \angle ACE$ $m \angle ABC = m \angle DBC + m \angle ABD$	3. Angle Addition Postulate.
4. $m \angle ACB > m \angle ABC$	4. Substitution Postulate.

Q Triangles ABC and DBC, as shown in the accompanying figure, are constructed in a way such that $m \angle DBC > m \angle ABC$ and $m \angle ABC > m \angle ACB$. Prove that $m \angle DBC > m \angle ACB$.

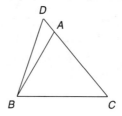

A This proof will involve applying the Transitive Property of Inequality, which states that if the first of three quantities is greater than the second and the second is greater than the third, then the first is greater than the third.

Statement	Reason
1. $m \angle DBC > m \angle ABC$	1. Given.
2. $m \angle ABC > m \angle ACB$	2. Given.
3. $m \angle DBC > m \angle ACB$	3. Transitive Property of Inequality.

Postulate 7

If equal quantities are subtracted from unequal quantities, the differences are unequal and in the same order.

(If $b > c$, then $b - a > c - a$.)

Postulate 8

If unequal quantities are subtracted from equal quantities, the differences are unequal and in the opposite order.

(If $a < b$ and $c = d$, then $c - a > d - b$.)

Postulate 9

If both sides of an inequality are multiplied by a positive number, the products are unequal and in the same order.

(If $a < b$ and c is a positive number, then $ac < bc$.)

Postulate 10

If both sides of an inequality are multiplied by a negative number, the products are unequal in the opposite order.

(If $a < b$ and c is a negative number, then $ac > bc$.)

Postulate 11

If unequal quantities are divided by equal positive quantities, the quotients are unequal in the same order.

(If $a > b$, c and d are positive, and $c = d$, then $a \div c > b \div d$.)

Postulate 12

If unequal quantities are divided by equal negative quantities, the quotients are unequal in the opposite order.

(If $a > b$, c and d are negative and $c = d$, then $a \div c < b \div d$.)

Postulate 13

Given real numbers a, and b, exactly one of the following is true: $a < b$, $a = b$, or $a > b$. This is commonly referred to as the Uniqueness of Order Postulate, or The Trichotomy Postulate.

Theorem 1

The shortest segment joining a line and a point outside the line is the perpendicular segment from the point to the line.

Theorem 2

For any real numbers a, b and c, if $c = a + b$ and $a > o$, then $c > b$.

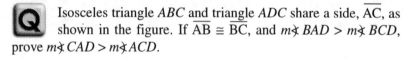

Problem Solving Examples:

Q Isosceles triangle ABC and triangle ADC share a side, $\overline{AC}$, as shown in the figure. If $\overline{AB} \cong \overline{BC}$, and $m \measuredangle BAD > m \measuredangle BCD$, prove $m \measuredangle CAD > m \measuredangle ACD$.

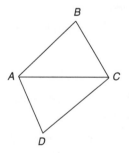

A By the Isosceles Triangle Theorem, we conclude that the base angles of isosceles triangle ABC are equal in measure. We will then subtract these equal angles from the given unequal angles and conclude, by the Subtraction Postulate of Inequality, that the differences are unequal in the same order.

Statement	Reason
1. $m \sphericalangle BAD > m \sphericalangle BCD$	1. Given.
2. $\overline{AB} \cong \overline{BC}$	2. Given.
3. $m \sphericalangle BAC = m \sphericalangle BCA$	3. If two sides of a triangle are congruent, the angles opposite these sides are of equal measure.
4. $m \sphericalangle BAD - m \sphericalangle BAC > m \sphericalangle BCD - m \sphericalangle BCA$	4. If equal quantities are subtracted from unequal quantities, the differences are unequal in the same order. (ee steps (1) and (3))
5. $m \sphericalangle CAD = m \sphericalangle BAD - m \sphericalangle BAC$ $m \sphericalangle ACD = m \sphericalangle BCD - m \sphericalangle BCA$	5. Angle Addition Postulate.
6. $m \sphericalangle CAD > m \sphericalangle ACD$	6. Substitution Postulate.

 In the figure, we are given $\sphericalangle DEF$ and $\sphericalangle ABC$, $m \sphericalangle DEF = m \sphericalangle ABC$. If $m \sphericalangle DEH > m \sphericalangle ABG$, then prove that $m \sphericalangle HEF < m \sphericalangle GBC$.

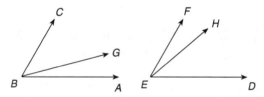

 A postulate tells us that when unequal quantities are subtracted from equal quantities, the differences are unequal in the opposite order. For example, if $10 = 10$ and $8 > 4$, then $10 - 8 < 10 - 4$ or $2 < 6$. This postulate will be utilized here.

Statement	Reason
1. $m \sphericalangle DEF = m \sphericalangle ABC$	1. Given.
2. $m \sphericalangle DEH > m \sphericalangle ABG$	2. Given.
3. $m \sphericalangle DEF - m \sphericalangle DEH < m \sphericalangle ABC - m \sphericalangle ABG$ or $m \sphericalangle HEF < m \sphericalangle GBC$	3. If unequal quantities are subtracted from equal quantities, the differences are unequal in the opposite order.

7.2   Inequalities in Triangles

Theorem 1

The sum of the lengths of two sides of a triangle is greater than the length of the third side.

Theorem 2

The measures of an exterior angle of a triangle is greater than the measure of either non-adjacent interior angle.

Theorem 3

If two sides of a triangle are unequal, the angles opposite these sides are unequal and the greater angle lies opposite the greater side.

Theorem 4

If two angles of a triangle are unequal, the sides opposite these angles are unequal and the greater side lies opposite the greater angle.

Theorem 5

If two sides of one triangle are equal to two sides of a second triangle and the included angle of the first is greater than the included angle of the second, then the third side of the first triangle is greater than the third side of the second.

Theorem 6

If two sides of one triangle are equal to two sides of a second triangle and the third side of the first is greater than the third side of the second, then the angle opposite the third side of the first triangle is greater than the angle opposite the third side of the second.

Problem Solving Examples:

Q Given: Point P is an interior point of $\triangle ABC$.
Prove: $AB + AC > BP + PC$ (Hint: extend $\overleftrightarrow{BP}$ so that it inter-sects $\overline{AC}$ at point N).

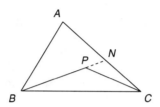

A We are asked to prove an inequality involving the sides of tri-angles. Therefore, we apply the Triangle Inequality Theorem. This theorem states that the sum of the lengths of any two sides of a triangle is greater than the length of the third. We must now find tri-angles whose sides include $\overline{AB}$, $\overline{AC}$, $\overline{BP}$, and $\overline{PC}$. We use the hint and extend $\overleftrightarrow{BP}$ so that $\overleftrightarrow{BP}$ intersects $\overline{AC}$ at point N. Thus, we obtain $\triangle ABN$ and $\triangle PNC$ whose sides comprise all the lengths in question—plus an extra length PN. By the Triangle Inequality Theorem, we know in $\triangle ABN$ that $\overline{AB} + \overline{AN} > \overline{BN}$ or (since $\overline{BN} = \overline{BP} + \overline{PN}$) that $\overline{AB} + \overline{AN} > \overline{BP} + \overline{PN}$. In $\triangle PNC$, $\overline{PN} + \overline{NC} > \overline{PC}$. Summing the two equations together, we obtain an inequality involving $\overline{AB}$, $\overline{AC}$, $\overline{BP}$, and $\overline{PC}$: $\overline{AB} + \overline{AN} + \overline{NC} + \overline{PN} > \overline{BP} + \overline{PC} + \overline{PN}$. Combining $\overline{AN} + \overline{NC}$ to obtain $\overline{AC}$, and cancelling $\overline{PN}$ from both sides, we obtain the desired result: $\overline{AB} + \overline{AC} > \overline{BP} + \overline{PC}$.

Statement	Reason
1. Point P is in the interior of $\triangle ABC$	1. Given.
2. $\overleftrightarrow{BP}$ intersects $\overline{AC}$ at point N	2. Two noncoincident, nonparallel coplanar lines intersect at a point.
3. $\overline{AB} + \overline{AN} > \overline{BN}$ $\overline{PN} + \overline{NC} > \overline{PC}$	3. Triangle Inequality Theorem.
4. $\overline{BN} = \overline{PN} + \overline{PB}$ $\overline{AC} = \overline{AN} + \overline{NC}$	4. Point Betweenness Postulate.
5. $\overline{AB} + \overline{AN} + \overline{PN} + \overline{NC} > \overline{BN} + \overline{PC}$	5. Addition Postulate of Inequality.
6. $\overline{AB} + \overline{AC} + \overline{PN} > \overline{PN} + \overline{PB} + \overline{PC}$	6. Substitution Postulate.
7. $\overline{AB} + \overline{AC} > \overline{PB} + \overline{PC}$	7. If $a > b$, then $a - c > b - c$.

Q $\triangle ABC$ is drawn in the accompanying figure with $BC > BA$. If $\overline{CD}$ bisects $\sphericalangle BCA$ and $\overline{AE}$ bisects $\sphericalangle BAC$, prove that $m\sphericalangle EAC > m\sphericalangle DCA$.

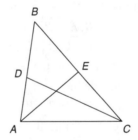

A By the definition of angle bisector, $m\sphericalangle EAC = \frac{1}{2}m\sphericalangle BAC$ and $m\sphericalangle DCA = \frac{1}{2}m\sphericalangle BCA$. Therefore, to show $m\sphericalangle EAC > m\sphericalangle DCA$, we show $m\sphericalangle BAC > m\sphericalangle BCA$. To accomplish this, we use the theorem that states: if two sides of a triangle are unequal in length then the opposite angles are unequal in measure, and the greater angle lies opposite the greater side.

Statement	Reason
1. In $\triangle ABC$, $\overline{BC} > \overline{BA}$	1. Given.
2. $m \sphericalangle BAC > m \sphericalangle BCA$	2. If two sides of a triangle are unequal, the angles opposite these sides are unequal, and the greater angle lies opposite the greater side.
3. $\overline{AE}$ bisects $\sphericalangle BAC$ $\overline{CD}$ bisects $\sphericalangle BCA$	3. Given.
4. $m \sphericalangle EAC = \frac{1}{2} m \sphericalangle BAC$ $m \sphericalangle DCA = \frac{1}{2} m \sphericalangle BCA$	4. A bisector of an angle divides the angle into two congruent angles.
5. $m \sphericalangle ECA > m \sphericalangle DCA$	5. Halves of unequal quantities are unequal in the same order. Also, Substitution Postulate.

 Given: $\triangle ABC$; D is a point between A and C; $\overline{BD} > \overline{AB}$
Prove: $\overline{BC} > \overline{AB}$

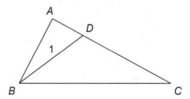

 $\overline{BC}$ and $\overline{AB}$ are sides of $\triangle ABC$. $\overline{BC} > \overline{AB}$ only if the angle opposite $\overline{BC}$ is greater in measure than the angle opposite $\overline{AB}$ – that is, $m \sphericalangle A > m \sphericalangle C$.

To show $m \sphericalangle A > m \sphericalangle C$, we relate both to $m \sphericalangle 1$. Note in $\triangle ABD$, $\overline{BD} > \overline{AB}$. This implies $m \sphericalangle A > m \sphericalangle 1$. Furthermore, in $\triangle BDC$, $\sphericalangle 1$ is an exterior angle while $\sphericalangle C$ is a remote interior angle. Thus, $m \sphericalangle 1 > m \sphericalangle C$.

Combining $m\angle A > m\angle 1$ and $m\angle 1 > m\angle C$, we obtain $m\angle A > m\angle$ C; and, therefore, $BC > AB$.

Statement	Reason
1. $\triangle ABC$; D is a point between A and C; $\overline{BD} > \overline{AB}$	1. Given.
2. $m\angle A > m\angle 1$	2. If two sides of a triangle are not congruent, the angle with the greater measure is opposite the longer side.
3. $m\angle 1 > m\angle C$	3. The measure of an exterior angle of a triangle is greater than the measure of either remote interior angle.
4. $m\angle A > m\angle C$	4. Transitive property of inequalities.
5. $\overline{BC} > \overline{AB}$	5. If two angles of a triangle are not congruent, the longer side is opposite the angle with the greater measure.

7.3   Inequalities in Circles

Theorem 1

In the same circle or in equal circles, if two central angles are unequal, then their arcs are unequal, and the greater angle has the greater arc.

Theorem 2

In the same circle, or in equal circles, if two arcs are unequal, then their central angles are unequal, and the greater arc has the greater central angle.

Theorem 3

In the same circle, or in equal circles, if two chords are unequal, then they are at unequal distances from the center. The longer a chord is, the smaller is the distance from the center.

If $\overline{CD} > \overline{AB}$, $\overline{OE} \perp \overline{AB}$, and $\overline{OE} \perp \overline{CD}$, then $\overline{OF} < \overline{OE}$

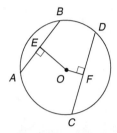

Problem Solving Example:

 In circle O, $\overline{BD} > \overline{AC}$. Prove that $\overline{CD} > \overline{AB}$.

 In a circle, if two chords are unequal, then their minor arcs are unequal in measure and the greater chord has the greater minor arc.

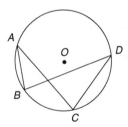

Since $\overline{BD} > \overline{AC}$, $m\overarc{BCD} > m\overarc{ABC}$. By reflexivity, $\overarc{BC} \cong \overarc{BC}$.

Hence, $m\overarc{BCD} - m\overarc{BC} > m\overarc{ABC} - m\overarc{BC}$ or $m\overarc{CD} > \overarc{AB}$. (This follows due to the fact when equal quantities are subtracted from unequal quantities, the results are unequal in the same order.)

If minor arcs of the same circle are unequal, then their chords are unequal and the greater minor arc has the greatest chord. Hence, since $m\overarc{CD} > m\overarc{AB}$, $\overline{CD} > \overline{AB}$.

Quiz: Triangles and Congruent Triangles – Geometric Inequalities

1. What is the value of x?

   (A) 20°

   (B) 40°

   (C) 60°

   (D) 90°

   (E) 30°

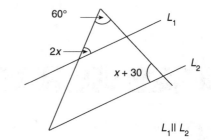

2. Find z.

   (A) 29°

   (B) 54°

   (C) 61°

   (D) 88°

   (E) 92°

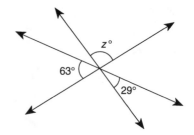

3. If $a \parallel b$ and $c \parallel d$, find $m \angle 5$.

   (A) 55°

   (B) 65°

   (C) 75°

   (D) 95°

   (E) 125°

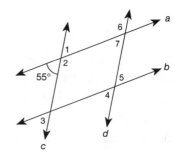

4. △*MNO* is isosceles. If the vertex angle, ∡*N*, has a measure of 96°, find the measure of ∡*M*.

   (A) 21°

   (B) 42°

   (C) 64°

   (D) 84°

   (E) 96°

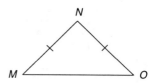

5. Find the area of rectangle *UVXY*.

   (A) 17 cm²

   (B) 34 cm²

   (C) 35 cm²

   (D) 70 cm²

   (E) 140 cm²

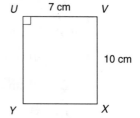

6. Find the length of $\overline{BO}$ in rectangle *BCDE* if the diagonal $\overline{EC}$ is 17 mm.

   (A) 6.55 mm

   (B) 8 mm

   (C) 8.5 mm

   (D) 17 mm

   (E) 34 mm

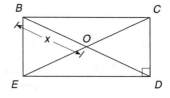

7. In rhombus *DEFG*, DE = 7 cm. Find the perimeter of the rhombus.

    (A)   14 cm

    (B)   28 cm

    (C)   42 cm

    (D)   49 cm

    (E)   56 cm

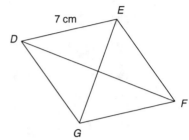

8. *ABCD* is an isosceles trapezoid. Find the perimeter.

    (A)   21 cm

    (B)   27 cm

    (C)   30 cm

    (D)   50 cm

    (E)   54 cm

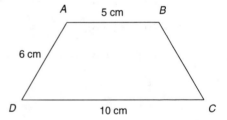

9. In △*ABC*, *AD* is drawn so that ∡3 = ∡4. All of the following are true EXCEPT

    (A) $\overline{AB} > \overline{BD}$.

    (B) $\overline{BD} < \overline{AB}$.

    (C) $\overline{BD} = \overline{BC}$.

    (D)   The measure of ∡4 is greater than the measure of ∡1.

    (E) $\overline{BD} > \overline{AB}$.

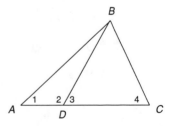

10. If, in the triangle shown here, $\overline{AS} < \overline{ST}$ which of the following *cannot* be the value of t.

   (A) 20

   (B) 36

   (C) 65

   (D) 71

   (E) 80

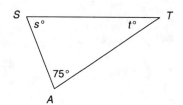

ANSWER KEY

1.	(E)		6.	(C)
2.	(D)		7.	(B)
3.	(A)		8.	(B)
4.	(B)		9.	(E)
5.	(D)		10.	(E)

CHAPTER 8

Geometric Proportions and Similarity

8.1   Ratio and Proportion

Definition 1

A ratio is the comparison of one number to another, expressed in quotient form. A ratio is therefore a fraction, with a denominator not equal to zero. The ratio of a to b ($b \neq 0$) is represented by a/b.

Definition 2

A proportion is a statement that equates two ratios.

Definition 3

In the proportion $a/b = c/d$, the numbers a and d are called the extremes of the proportion, and the numbers b and c are called the means of the proportion. The single term, d, is called the fourth proportional.

Problem Solving Examples:

 Is $\frac{12}{20} = \frac{36}{60}$ a proportion?

 A proportion is an equation which states that two ratios are equal.

Since $\frac{12}{20} = \frac{3}{5}$ and $\frac{36}{60} = \frac{3}{5}$, the ratios $\frac{12}{20}$ and $\frac{36}{60}$ are equal, and, therefore, $\frac{12}{20} = \frac{36}{60}$ is a proportion.

 Solve for the unknown, c, in the proportion $18 : 6 = c : 9$.

 A theorem tells us that, in a proportion, the product of the means is equal to the product of the extremes.

In this problem, 6 and c are the means and 18 and 9 are the extremes. Therefore,

$$6c = 18 \times 9$$
$$6c = 162$$
$$c = \frac{162}{6} = 27.$$

The answer $c = 27$ can be checked by substituting back into the original proportion.

$$18 : 6 = 27 : 9$$
$$3 : 1 = 3 : 1$$

Since the two ratios are equal, $c = 27$.

Theorem 1

In a proportion, the product of the means is equal to the product of the extremes.

(If $a/b = c/d$, then $bc = ad$.)

Theorem 2

A proportion may be written by inversion.

(If $a/b = c/d$, then $b/a = d/c$.)

Theorem 3

The means may be interchanged in any proportion.

(If $a/b = c/d$, then $a/c = b/d$.)

Theorem 4

The extremes may be interchanged in any proportion.

(If $a/b = c/d$, then $d/b = c/a$.)

Theorem 5

A proportion may be written by addition.

(If $\dfrac{a}{b} = \dfrac{c}{d}$, then $\dfrac{a+b}{b} = \dfrac{c+d}{d}$.)

Theorem 6

A proportion may be written by subtraction.

(If $\dfrac{a}{b} = \dfrac{c}{d}$, then $\dfrac{a-b}{b} = \dfrac{c-d}{d}$.)

Theorem 7

If three terms of one proportion are equal, respectively, to three terms of a second proportion, the fourth terms are equal.

Also, (if $\dfrac{a}{b} = \dfrac{c}{d}$, and $\dfrac{a}{b} = \dfrac{c}{e}$, then $d = e$.)

Theorem 8

If the numerators of a proportion are equal, then the denominators are equal.

(If $a/b = c/d$ and $a = c$, then $b = d$.)

Theorem 9

Given a proportion, the ratio of the sum of the numerators to the sum of the denominators forms a proportion with either of the original ratios.

(If $\dfrac{a}{b} = \dfrac{c}{d}$, then $\dfrac{a+c}{b+d} = \dfrac{a}{b}$, and $\dfrac{a+c}{b+d} = \dfrac{c}{d}$.)

Theorem 10

If the product of two numbers (not zero) is equal to the product of two other numbers (not zero), either pair of numbers may be made the means and the other pair may be made the extremes in a proportion.

(If $ab = cd \neq 0$, then $a/c = d/b$ and $b/c = d/a$.

Also, $c/a = b/d$ and $d/a = b/c$.)

Problem Solving Example:

Find the fourth proportional to 3, 4, and 9.

Let x = the fourth proportional to 3, 4 and 9.

If x is the fourth proportional, then the equation $3 : 4 = 9 : x$ is a proportion. Since the product of the means equals the product of the extremes, we have

$$3x = 9 \times 4$$
$$3x = 36$$
$$x = 12.$$

To check this, we substitute $x = 12$ into $3 : 4 = 9 : x$ and get $3 : 4 = 9 : 12$, which reduces to $3 : 4 = 3 : 4$, a proportion.

Therefore, the fourth proportional to 3, 4, and 9, is 12.

8.2   Parallel Lines and Proportions

Theorem 1

A line parallel to one side of a triangle divides the other two sides proportionally.

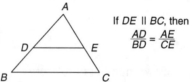

If *DE* ∥ *BC*, then
$$\frac{AD}{BD} = \frac{AE}{CE}$$

Theorem 2

If a line divides two sides of a triangle proportionally, it is parallel to the third side.

Theorem 3

The bisector of one angle of a triangle divides the opposite side in the same ratio as the other two sides.

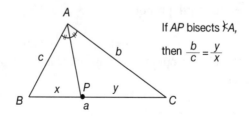

If *AP* bisects ⦨*A*,

then $\dfrac{b}{c} = \dfrac{y}{x}$

Theorem 4

Three or more parallel lines intercept proportional segments on any two transversals.

If $\ell_1 \parallel \ell_2 \parallel \ell_3$, and ℓ_x and ℓ_y are transversals intersecting the parallel lines at points A through F as shown, then $\dfrac{AB}{BC} = \dfrac{FE}{DE}$

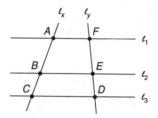

Problem Solving Example:

Q Show that the medians of a triangle are concurrent at a point on each median located two-thirds of the way from each vertex to the opposite side.

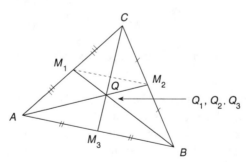

A Referring to the figure above, we must show (1) Q is a point on $\overline{AM_2}$, $\overline{BM_1}$, and $\overline{CM_3}$; and (2) that $AQ = \frac{2}{3} AM_2$, $BQ = \frac{2}{3} BM_1$, and $CQ = \frac{2}{3} CM_3$. Since $AM_2 = AQ + QM_2$, $QM_2 = AM_2 - AQ$

$= AM_2 - \frac{2}{3} AM_2 = \frac{1}{3} AM_2$, or $QM_2 = \frac{1}{3} AM_2$. Thus, to prove the second part, we need to show that the point of concurrency divides the median AM_2 into two segments such that the segment near the vertex is twice the length of the segment near the side. In algebraic notation, what we must show for the second part is that

$$\frac{AQ}{QM_2} = \frac{BQ}{QM_1} = \frac{CQ}{QM_3} = 2$$

Wherever there are proportions, it is wise to look for similar triangles. Suppose we first wish to show that

$$\frac{AQ}{QM_2} = \frac{2}{1}.$$

Then, we find two triangles such that $\overline{AQ}$ and $\overline{QM_2}$ are corresponding sides. $\triangle AQB$ and $\triangle M_2 QM_1$ are such triangles. We show similarity by the *A-A* Similarity Theorem. Then

$$\frac{AQ}{QM_2} = \frac{BQ}{QM_1} = \frac{AB}{M_1 M_2} \times \frac{AB}{M_1 M_2}$$

is known. $\overline{M_1 M_2}$ is a midline of $\triangle ABC$. Therefore,

$$AB = 2 \times M_1 M_2, \quad \frac{AB}{M_1 M_2} = 2, \text{ and } \frac{AQ}{QM_2} = 2,$$

proving the second part. We can repeat the procedure for the two other sides.

NOTE: It may seem that we have proven the second part without actually showing that the lines are indeed concurrent, but we really don't need the concurrency to show the two-thirds division. In fact, the exact reverse is true. We show that the intersection of any two medians is a point two-thirds the length of each median from the respective vertex. Since there is only one point that is two-thirds the median length from the vertex, the points of intersection must be the same. Concurrency (part one) follows from the two-thirds division (part two).

Given: $\overline{AM_2}$, $\overline{BM_1}$, and $\overline{CM_3}$ are medians of $\triangle ABC$.

Prove: (1) Q is a point on $\overline{AM_2}$, $\overline{BM_1}$, and $\overline{CM_3}$.

(2) $AQ = \frac{2}{3} AM_2$, $BQ = \frac{2}{3} BM_1$, $CQ = \frac{2}{3} CM_3$.

Statement	Reason
1. $\overline{AM}_2$, $\overline{BM}_1$, and $\overline{CM}_3$ are medians of $\triangle ABC$	1. Given.
2. Q_1 is the intersection of $\overline{AM}_2$ and $\overline{BM}_1$.	2. Two non-parallel, non-coincident lines intersect in a point.
3. $\measuredangle AQ_1B \cong \measuredangle M_2Q_1M_1$	3. Opposite angles formed by two intersecting lines are congruent.
4. $\overline{M_1M_2}$ is a midline of $\triangle ABC$	4. The segment that connects the midpoints of two sides of a triangle is a midline of the triangle.
5. $\overline{M_1M_2} \parallel \overline{AB}$	5. The midline of a triangle is parallel to the third side.
6. $\measuredangle M_1M_2Q \cong \measuredangle QAB$	6. Alternate interior angles of a transversal are congruent.
7. $\triangle M_1QM_2 \sim \triangle AQB$	7. If two angles of one triangle are congruent with the corresponding two angles of a second triangle, then the two triangles are similar.
8. $\dfrac{M_2Q}{AQ} = \dfrac{M_1Q}{BQ} = \dfrac{M_1M_2}{AB}$	8. The corresponding sides of similar triangles are proportional.
9. $AB = 2 \times M_1M_2$ or $\dfrac{M_1M_2}{AB} = \dfrac{1}{2}$	9. The midline of a triangle is half the length of the third side.
10. $\dfrac{M_2Q}{AQ} = \dfrac{M_1Q}{BQ} = \dfrac{1}{2}$	10. Substitution Postulate.
11. $AM_2 = AQ + M_2Q$ $BM_1 = BQ + M_1Q$	11. Q_1 is the intersection of two segments AM_2 and BM_1. Therefore, Q_1 must lie between the endpoints of the segment. The equations at left follow from the definition of betweenness.

12. $M_2Q = \frac{1}{2} AQ_1$ $M_1Q_1 = \frac{1}{2} BQ_1$	12. Multiplication Postulate and Step 10.
13. $AM_2 = 1\frac{1}{2} AQ_1$ $BM_1 = 1\frac{1}{2} BQ_1$	13. Substitution of Step 12 into 11.
14. $AQ_1 = \frac{2}{3} AM_2$ $BQ_1 = \frac{2}{3} BM_1$	14. Multiplication Postulate.
15. Q_2 is the intersection of $\overline{AM}_2$ and $\overline{CM}_3$	15. Two non-parallel, non-coincident lines intersect in a unique point.
16. $AQ_2 = \frac{2}{3} AM_2$ $CQ_2 = \frac{2}{3} CM_3$	16. Obtained by repeating procedure used from Steps 3 through 13. The midline is now M_1M_3 not M_1M_2.
17. Q_3 is the intersection of medians $\overline{CM}_3$ and $\overline{BM}_1$	17. Two non-parallel non-coincident lines intersect in a unique point.
18. $CQ_3 = \frac{2}{3} CM_3$ $BQ_3 = \frac{2}{3} BM_1$	18. Obtained by repeating procedure used from Steps 3 through 13. The midline is now M_1M_3.
19. Q_1 is a point on segment $\overline{AM}_2$ such that $AQ_1 = \frac{2}{3} AM_2$. Q_2 is a point on segment AM_2 such that $AQ_2 = \frac{2}{3} AM_2$.	19. Repetition of results from Steps 14 and 16.
20. Point Q_1 is point Q_2	20. On a given segment, there is only one point on the segment that is a given distance from a given endpoint.
21. Q_2 is a point on segment $\overline{CM}_3$ such that $CQ_2 = \frac{2}{3} CM_3$. Q_3 is a point on segment $\overline{CM}_3$ such that $CQ_3 = \frac{2}{3} CM_3$.	21. Repetition of results from Steps 16 and 18.
22. Point Q_2 is point Q_3	22. Same reason as Step 20.

23. Let point $Q = Q_1 = Q_2 = Q_3$

23. From Steps 20 and 22, we have shown $Q_1 = Q_2 = Q_3$ and here we give this common point of intersection of the three medians a more general name.

24. Q is a point on $\overline{AM}_2$, $\overline{BM}_1$, and $\overline{CM}_3$

24. Follows from Step 23 and the definitions of C_1, C_2, and C_3.

25. $AQ = \frac{2}{3} AM_2$
    $BQ = \frac{2}{3} BM_1$
    $CQ = \frac{2}{3} CM_3$

25. Substitution Postulate (Substitute Q for Q_1, Q_2, and Q_3 in Steps 14, 16, and 18.)

8.3   Similar Triangles

Definition 1

Two polygons are similar if their corresponding angles are equal and their corresponding sides are in proportion. The symbol for similarity is ~; thus, $\triangle ABC$ is similar to $\triangle DEF$ is written $\triangle ABC \sim \triangle DEF$.

Definition 2

The ratio of similitude refers to the common ratio of corresponding sides of similar polygons.

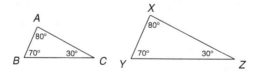

$\triangle ABC \sim \triangle XYZ$; $\angle A = \angle X$, $\angle B = \angle Y$, $\angle C = \angle Z$ as shown,

therefore $\dfrac{AB}{XY} = \dfrac{AC}{XZ} = \dfrac{BC}{YZ}$

Postulate

Two triangles are similar if and only if two angles of one triangle are equal to two angles of the other triangle.

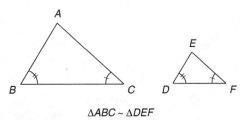

$\triangle ABC \sim \triangle DEF$

Problem Solving Example:

Q Triangles ABC and $A'B'C'$ have dimensions as indicated in the figure. Angle A = angle A', angle B = angle B'. Find the measures of sides $A'B'$ and $C'B'$.

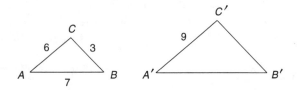

A Two triangles are similar if, and only if, two angles of one triangle are equal in measure to two angles of the other triangle. Since $m \angle A = m \angle A'$, and $m \angle B = m \angle B'$, the two triangles $\triangle ABC$ and $\triangle A'B'C'$ are similar. In similar triangles, corresponding sides are proportional. Hence,

$$\frac{AC}{A'C'} = \frac{CB}{C'B'} = \frac{AB}{A'B'}.$$

Substituting the given,

$$\frac{6}{9} = \frac{3}{C'B'} = \frac{7}{A'B'}$$

We can first solve for $C'B'$, using the proportion:

$$\frac{6}{9} = \frac{3}{C'B'}$$
$$6 \cdot C'B' = 27$$
$$C'B' = \frac{9}{2}$$

We then solve for $A'B'$, using the proportion:

$$\frac{6}{9} = \frac{7}{A'B'}$$
$$6 \cdot A'B' = 63$$
$$A'B' = \frac{21}{2}$$

Therefore, side $A'B' = \dfrac{21}{2}$, and side $C'B' = \dfrac{9}{2}$.

Theorem 1

All congruent triangles are similar.

Theorem 2

Two triangles similar to the same triangle are similar to each other.

Theorem 3

If one triangle has an angle equal to that of another triangle, and the respective sides including these angles are proportionate, the triangles are similar.

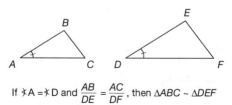

If $\angle A = \angle D$ and $\dfrac{AB}{DE} = \dfrac{AC}{DF}$, then $\triangle ABC \sim \triangle DEF$

Theorem 4

If three sides of one triangle are in proportion to the three corresponding sides of a second triangle, the triangles are similar. (S.S.S., Similarity Theorem)

Theorem 5

If three angles of one triangle are congruent to three corresponding angles of another triangle, then the two triangles are similar. (A.A.A., Similarity Theorem)

Problem Solving Examples:

 A right triangle has legs of length 6 and 8 inches. $\overline{CD}$ bisects the right angle. Find the lengths of $\overline{AD}$ and $\overline{DB}$.

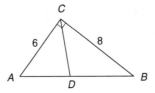

 We must somehow relate the unknown segment lengths to the given data to derive the required results.

Recall that the bisector of one angle of a triangle divides the opposite side so that the lengths of its segments are proportional to the lengths of the adjacent sides. Thus,

$$\frac{6}{8} = \frac{AD}{DB}.$$

We see that $\frac{8}{8} = \frac{DB}{DB}$ is always true. Hence, adding this to the above proportion, we obtain

$$\frac{6+8}{8} = \frac{AD+DB}{DB}.$$

But, $AD + DB = AB$, the hypotenuse of a right triangle. Applying the Pythagorean theorem, $a^2 + b^2 = c^2$, where $a = 6$, $b = 8$, we obtain

$$6^2 + 8^2 = c^2$$

$$36 + 64 = c^2$$

$$100 = c^2.$$

Thus, $c = 10$, or hypotenuse $AB = 10$. Substituting, we obtain

$$\frac{6+8}{8} = \frac{AD+DB}{DB}$$

$$\frac{14}{8} = \frac{AB}{DB}$$

$$\frac{14}{8} = \frac{10}{DB}.$$

Since, in a proportion, the product of the means is equal to the product of the extremes, we obtain

$$14 \times DB = 8 \cdot 10,$$

$$DB = \frac{80}{14} = \frac{40}{7}.$$

To find AD, we notice that

$$AD = AB - DB$$

$$= 10 - \frac{40}{7} = \frac{30}{7}.$$

Therefore,

$$AD = \frac{30}{7} \text{ and } DB = \frac{40}{7}.$$

Given the A.A.A. (Angle, Angle, Angle) Similarity Theorem, prove the A.A. (Angle, Angle) Similarity Theorem.

The A.A.A. Similarity Theorem states: If there exists a correspondence between $\triangle ABC$ and $\triangle DEF$ such that corresponding angles are congruent, then $\triangle ABC \sim \triangle DEF$.

Suppose we are given two triangles and that the corresponding angles of two pairs of angles are congruent. Noting that the sum of the angles of any triangle is 180°, it follows that the corresponding angles of the third pair of angles are also congruent and that the two triangles are similar by the A.A.A. theorem. Hence, we obtain a generalization of the A.A.A. Similarity Theorem. The A.A. Similarity Theorem: If there exists a correspondence between $\triangle ABC$ and $\triangle DEF$ such that two angles of $\triangle ABC$ are congruent to the corresponding angles of $\triangle DEF$, then $\triangle ABC \sim \triangle DEF$.

Theorem 6

If two triangles are similar, the measures of corresponding altitudes have the same ratio as the measures of any two corresponding sides of the triangles.

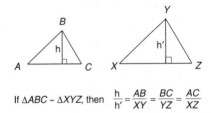

If $\triangle ABC \sim \triangle XYZ$, then $\dfrac{h}{h'} = \dfrac{AB}{XY} = \dfrac{BC}{YZ} = \dfrac{AC}{XZ}$

Theorem 7

The perimeters of two similar triangles have the same ratio as the measures of any pair of corresponding sides of the triangles.

Theorem 8

If a given triangle is similar to a triangle that is congruent to a third triangle, then the given triangle is similar to the third triangle.

Theorem 9

Given $\triangle ABC$; if $\dfrac{AB}{AD} = \dfrac{AC}{AE}$, then $\dfrac{AB}{AD} = \dfrac{BC}{DE}$.

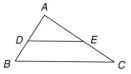

Theorem 10

The ratio of similitude of any pair of similar triangles equals the square root of the ratio of their areas.

Theorem 11

The ratio of the areas of any two similar triangles is equal to the ratio of the squares of the lengths of any two corresponding sides, or any two corresponding line segments of the two similar triangles.

Problem Solving Examples:

Q In the accompanying figure, $\triangle ABC$ is isosceles with $\overline{AB} \cong \overline{AC}$. Segment AF is the altitude on $\overline{BC}$. From a point on AB, call it D, a perpendicular is drawn which is extended to meet $\overleftrightarrow{BC}$. It meets $\overline{BC}$ at point P. Prove that $FC : DB = AC : PB$.

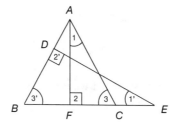

A $\overline{FC}$ and $\overline{AC}$ are sides of ΔFCA and correspond to sides $\overline{DB}$ and $\overline{PB}$ of ΔDBP. By proving $\Delta FCA \sim \Delta DBP$, we can conclude that $\overline{FC} : \overline{DB} = \overline{AC} : \overline{PB}$ is a proportion, because there is a theorem which states that corresponding sides of similar triangles are proportional.

The corresponding angles of ΔFCA and ΔDBP have been numbered, in the diagram, by a "prime and no prime" system.

Similarity will be proved by showing two pairs of corresponding angles congruent.

Statement	Reason
1. $\overline{AB} = \overline{AC}$	1. Given.
2. $\angle 3 \cong \angle 3'$	2. If two sides of a triangle are congruent, then the angles opposite these sides are congruent.
3. $\overline{AF} \perp \overline{BC}$	3. An altitude drawn to a side is perpendicular to that side.
4. $\overline{PD} \perp \overline{AB}$	4. Given.
5. $\angle 2$ and $\angle 2'$ are right angles	5. Perpendicular lines intersect forming right angles.
6. $\angle 2 \cong \angle 2'$	6. All right angles are congruent.
7. $\Delta FCA \sim \Delta DBP$	7. A.A. $\cong$ A.A.
8. $\dfrac{FC \ (\text{opp. } \angle 1)}{DB \ (\text{opp. } \angle 1')} = \dfrac{AC \ (\text{opp. } \angle 2)}{PB \ (\text{opp. } \angle 2')}$	8. Corresponding sides of similar triangles are proportional.

Q   The sides of triangle ABC measure 5, 7, and 9. The shortest side of a similar triangle, $A'B'C'$, measures 10. (a) Find the measure of the longest side of triangle $A'B'C'$. (b) Find the ratio of the measures of a pair of corresponding altitudes in triangles ABC and $A'B'C'$. (c) Find the perimeter of triangle $A'B'C'$.

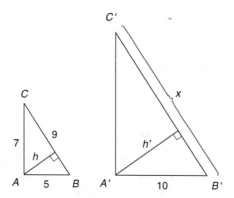

A (a) As seen in the diagram, the longest side of $\triangle A'B'C'$ is $\overline{B'C'}$; it corresponds to $\overline{BC}$, the longest side of $\triangle ABC$. The shortest corresponding sides are $\overline{AB}$ and $\overline{A'B'}$.

Since $\triangle ABC \sim \triangle A'B'C'$, a proportion exists between corresponding sides. For these triangles, one proportion is

$$\frac{AB}{A'B'} = \frac{BC}{B'C'}.$$

The only length in this proportion which is unknown is $B'C'$, and this can be determined by substitution and algebra.

Substituting:

$$\frac{5}{9} = \frac{10}{B'C'}$$

$$90 = 5\,B'C'$$

$$B'C' = 18.$$

The longest side of $\triangle A'B'C'$, $\overline{B'C'}$ measures 18.

(b) The ratio of the measures of the pair of altitudes, h and h' will be the same for these two similar triangles as will be the ratio of the measures of any other pair of corresponding linear parts.

Choose AB and $A'B'$ as the determinants of the ratio. Therefore,

$$\frac{h}{h'} = \frac{AB}{A'B'}.$$

By substitution, we conclude that

$$\frac{h}{h'} = \frac{5}{10} \text{ or } \frac{1}{2}.$$

(c) The perimeter of $A'B'C'$ is a non-angular measure and, as with the altitudes, will have a ratio of similitude which is the same as that possessed by all other pairs of linear corresponding parts. If p is the perimeter of $\triangle ABC$, and p' that of $\triangle A'B'C'$, we have

$$p : p' = 1 : 2.$$

By substitution, we obtain

$$21 : p' = 1 : 2, \text{ or } p' = 42.$$

Corollary 1

If two triangles are similar, then their corresponding sides are in proportion.

Corollary 2

The measures of any two corresponding line segments of two similar triangles have the same ratio as the measures of any pair of corresponding sides.

Corollary 3

If a line parallel to one side of a triangle intersects the other two sides, then it cuts off a triangle similar to the original triangle.

Corollary 4

Two triangles which are similar to the same triangle are similar to each other. If $\triangle ABC \sim \triangle DEF$, and $\triangle GHI \sim \triangle DEF$, then $\triangle ABC \sim \triangle GHI$.

Corollary 5

If the corresponding sides of two triangles are parallel to each other, the triangles are similar.

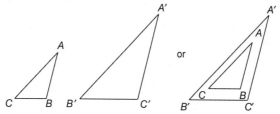

In both cases, *AB* ∥ *A′B′*, *BC* ∥ *B ′C′*, and *CA* ∥ *C′A′*; hence, △*ABC* ~ △*A′B′C′*

Corollary 6

If the corresponding sides of two triangles are perpendicular to each other, the triangles are similar.

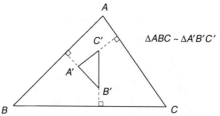

△*ABC* ~ △*A′B′C′*

Problem Solving Examples:

In the accompanying figure, the line segment, $\overline{KL}$, is drawn parallel to $\overline{ST}$, intersecting $\overline{RS}$ at *K* and $\overline{RT}$ at *L* in △*RST*. If $\overline{RK} = 5$, $\overline{KS} = 10$, and $\overline{RT} = 18$, then find $\overline{RL}$.

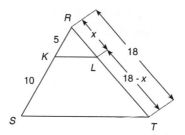

If a line is parallel to one side of a triangle, then it divides the other two sides proportionally. Since $\overline{KL} \parallel \overline{ST}$, $\overline{RT}$ and $\overline{RS}$ are divided proportionally.

Let $x = \overline{RL}$. Then $18 - x = \overline{LT}$. Set up the proportion

$$\frac{RK}{KS} = \frac{RL}{LT} \qquad \text{and substitute to obtain}$$

$$\frac{5}{10} = \frac{x}{18 - x}$$

or

$$10x = 90 - 5x$$

$$15x = 90$$

$$x = 6$$

Therefore, $x = \overline{RL} = 6$.

Alternatively, instead of forming a ratio of upper to lower segment, we can form a ratio of the upper segment to the whole side.

We are given $\overline{RT} = 18$, $\overline{RS} = 15$, and $\overline{RK} = 5$. If we let $\overline{RL} = x$, the proportion becomes

$$\frac{RK}{RS} = \frac{RL}{RT}$$

$$\frac{5}{15} = \frac{x}{18}$$

$$15x = 90$$

$$x = 6.$$

It is seen, then, that the same solution can be arrived at in several ways.

 In triangle ABC, $\overline{CD} = 6$, $\overline{DA} = 5$, $\overline{CE} = 12$, and $\overline{EB} = 10$, as shown in the figure. Is $\overline{DE}$ parallel to $\overline{AB}$?

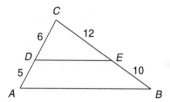

 A postulate tells us that if a line divides two sides of a triangle proportionally, the line is parallel to the third side.

Set up the ratio of upper to lower segments for both $\overline{CA}$ and $\overline{CB}$. If these ratios are equal, the lines are proportional and we can then conclude that $\overline{DE} \parallel \overline{AB}$.

$$\frac{CD}{DA} = \frac{6}{5} \quad \text{and} \quad \frac{CE}{EB} = \frac{12}{10} = \frac{6}{5}$$

Therefore, $\overline{CA}$ and $\overline{CB}$ are divided proportionally.

We can thus conclude that $\overline{DE}$ is parallel to $\overline{AB}$.

8.4   Properties of the Right Triangle and Similar Right Triangles

Theorem 1

The altitude on the hypotenuse of a right triangle is the mean proportional between the segments of the hypotenuse.

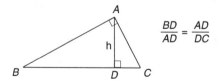

$$\frac{BD}{AD} = \frac{AD}{DC}$$

Theorem 2

In a right triangle $\triangle ABC$, the altitude on the hypotenuse separates the triangle into two triangles that are similar to each other and to the original triangle.

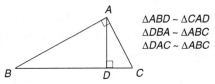

$\triangle ABD \sim \triangle CAD$
$\triangle DBA \sim \triangle ABC$
$\triangle DAC \sim \triangle ABC$

Theorem 3

Two right triangles are similar if the ratios of their hypotenuses and any pair of corresponding sides are proportional.

In $\triangle ABC$ and $\triangle CDE$ sharing right angle C, if $\dfrac{AC}{EC} = \dfrac{AB}{ED}$,

then $\triangle ABC \sim \triangle EDC$ and $\dfrac{AC}{EC} = \dfrac{BC}{DC}$

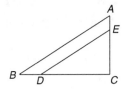

Theorem 4

The length of each leg of a given right triangle is the mean proportional between the length of the whole hypotenuse and the length of the projection of that leg on the hypotenuse.

$$\frac{BC}{AB} = \frac{AB}{BD} , \frac{BC}{AC} = \frac{AC}{DC}$$

Corollary

If acute angles of two right triangles are congruent, then the triangles are similar.

Problem Solving Examples:

Q In right triangle *ABC*, altitude $\overline{CD}$ is drawn to hypotenuse $\overline{AB}$, as shown in the diagram. If *CD* = 12 and *AD* exceeds *DB* by 7, find the lengths of *DB* and *AD*.

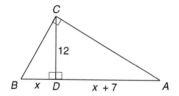

A By theorem, we know that the length of the altitude to the hypotenuse will be the mean proportional between the lengths of the segments of the hypotenuse. In this case, $\overline{AD} : \overline{CD} = \overline{CD} : \overline{BD}$.

If we let x = the length of $\overline{DB}$,

then $x + 7$ = the length of $\overline{AD}$.

By substituting this and *CD* = 12 we get $(x + 7) : 12 = 12 : x$. Since the product of the means equals the product of the extremes in a proportion, this equals $x^2 + 7x = 144$.

It follows then that $x^2 + 7x - 144 = 0$.

By factoring, this becomes $(x - 9)(x + 16) = 0$.

Therefore, $x = 9$ or $x = -16$. (The negative is rejected for lack of geometric significance.)

If $x = 9$, then $x + 7 = 16$.

Therefore, the length of $\overline{DB}$ = 9 and the length of $\overline{AD}$ = 16.

In right triangle ABC, altitude $\overline{CD}$ is drawn to hypotenuse $\overline{AB}$, as seen in the figure. If $AD = 6$ and $DB = 24$, find (a) CD and (b) AC.

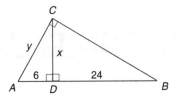

 If the altitude is drawn to the hypotenuse of a right triangle, as in $\triangle ACB$, then the length of the altitude (CD) is the mean proportional between the lengths of the segments of the hypotenuse ($\overline{AD}$ and $\overline{DB}$). Also, the triangles thus formed ($\triangle ADC$ and $\triangle CDB$) are both similar to the given triangle ($\triangle ACB$).

(a) The first part of the above conclusion allows us to form the proportion

$$\frac{AD}{CD} = \frac{CD}{DB}.$$

Since we want to determine CD, let $x =$ length of $\overline{CD}$ and substitute the given values of AD and DB into the proportion. This gives us $\frac{6}{x} = \frac{x}{24}$. The product of the means equals the product of the extremes, therefore, $x^2 = 144$ and $x = 12$; the length of altitude $\overline{CD}$ is 12.

(b) The second fact, that $\triangle ADC \sim \triangle ACB$, tells us that corresponding sides are in proportion. Therefore, to find AC, which is the hypotenuse of $\triangle ADC$ and the shorter leg of $\triangle ACB$, we set up the following proportion:

$$\frac{AD}{AC} = \frac{AC}{AB}.$$

If we let $y = AC$, and substitute in the given, we find that $\frac{6}{y} = \frac{y}{30}$. As stated in part (a), this expression is equal to $y^2 = 180$. Therefore,

$$y = \sqrt{180} = \sqrt{36} \cdot \sqrt{5} = 6\sqrt{5} \text{ which is the length of } AC.$$

8.5   Trigonometric Ratios

For definitions 1–4, refer to Figure 1. Given the right triangle $\triangle ABC$:

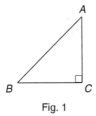

Fig. 1

Definition 1

$$\sin A = \frac{BC}{AB}$$

$$= \frac{\text{measure of side opposite } \angle A}{\text{measure of hypotenuse}}$$

Definition 2

$$\cos A = \frac{AC}{AB}$$

$$= \frac{\text{measure of side adjacent to } \angle A}{\text{measure of hypotenuse}}$$

Definition 3

$$\tan A = \frac{BC}{AC}$$

$$= \frac{\text{measure of side opposite } \angle A}{\text{measure of side adjacent to } \angle A}$$

Definition 4

$$\cot A = \frac{AC}{BC}$$

$$= \frac{\text{measure of side adjacent to } \sphericalangle A}{\text{measure of side opposite } \sphericalangle A}$$

Hint: To remember the ratios for sine, cosine and tangent, think of the acronym SOHCAHTOA.

S O H	C A H	T O A
sine = opposite ÷ hypotenuse	cosine = adjacent ÷ hypotenuse	tangent = opposite ÷ adjacent

cotangent is merely the inverse of tangent.

α	sin α	cos α	tan α	cot α
0°	0	1	0	undefined
30°	$\frac{1}{2}$	$\frac{\sqrt{3}}{2}$	$\frac{1}{\sqrt{3}} = \frac{\sqrt{3}}{3}$	$\sqrt{3}$
45°	$\frac{1}{\sqrt{2}} = \frac{\sqrt{2}}{2}$	$\frac{1}{\sqrt{2}} = \frac{\sqrt{2}}{2}$	1	1
60°	$\frac{\sqrt{3}}{2}$	$\frac{1}{2}$	$\sqrt{3}$	$\frac{1}{\sqrt{3}} = \frac{\sqrt{3}}{3}$
90°	1	0	undefined	0

Hint: For the ratios of 30° and 60° think of the 1, 2, $\sqrt{3}$ right triangle:

For the ratios of 45° think of the 1, 1, $\sqrt{2}$ right triangle:

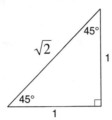

Hence all of the ratios follow.

Problem Solving Examples:

 Given the right triangle with $a = 3$, $b = 4$, and $c = 5$, find the values of the trigonometric functions of α.

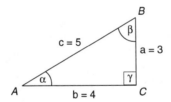

 In the accompanying figure, a is the side opposite angle α, b is the side opposite angle β, and c is the side opposite angle γ. The values of the trigonometric functions of α are:

$$\cos \alpha = \frac{\text{adjacent side,}}{\text{hypotenuse}} \qquad \sin \alpha = \frac{\text{opposite side,}}{\text{hypotenuse}}$$

$$\tan \alpha = \frac{\text{opposite side,}}{\text{adjacent side}} \qquad \cot \alpha = \frac{1}{\tan \alpha,}$$

$$\sec \alpha = \frac{1}{\cos \alpha,} \qquad \text{and} \quad \csc \alpha = \frac{1}{\sin \alpha.}$$

Therefore: $\cos \alpha = \dfrac{4}{5},$ $\sin \alpha = \dfrac{3}{5},$

$\tan \alpha = \dfrac{3}{4},$ $\cot \alpha = \dfrac{1}{3/4} = \dfrac{4}{3},$

$\sec \alpha = \dfrac{1}{4/5} = \dfrac{5}{4},$ $\csc \alpha = \dfrac{1}{3/5} = \dfrac{5}{3}.$

 Find the values of the six trigonometric functions of an angle, in a right triangle, whose opposite side is 3 and hypotenuse 5.

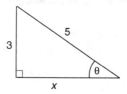

 We are given that, in a right triangle, the side opposite the angle is 3 and the hypotenuse is 5. To determine the adjacent side, x (see figure), we recall the Pythagorean Theorem which states that the sum of the square of the legs of a right triangle equals the square of the hypotenuse. Thus,

$$3^2 + x^2 = 5^2$$
$$9 + x^2 = 25$$
$$x^2 = 16$$
$$x = 4$$

Now that we know the value of each side of the triangle, we can find the values of the six trigonometric functions:

$$\sin \theta = \frac{\text{opposite side}}{\text{hypotenuse}} = \frac{3}{5}$$

$$\cos \theta = \frac{\text{adjacent side}}{\text{hypotenuse}} = \frac{4}{5}$$

$$\tan \theta \;=\; \frac{\text{opposite side}}{\text{adjacent side}} \;=\; \frac{3}{4}$$

$$\csc \theta \;=\; \frac{1}{\sin \theta} \;=\; \frac{\text{hypotenuse}}{\text{opposite side}} \;=\; \frac{5}{3}$$

$$\sec \theta \;=\; \frac{1}{\cos \theta} \;=\; \frac{\text{hypotenuse}}{\text{adjacent side}} \;=\; \frac{5}{4}$$

$$\cot \theta \;=\; \frac{1}{\tan \theta} \;=\; \frac{\text{adjacent side}}{\text{opposite side}} \;=\; \frac{4}{3}$$

8.6   Similar Polygons

Definition

Two polygons are similar if there is a one-to-one correspondence between their vertices such that all pairs of corresponding angles are congruent and the ratios of the measures of all pairs of corresponding sides are equal.

Theorem 1

The perimeters of two similar polygons have the same ratio as the measure of any pair of corresponding line segments of the polygons.

Theorem 2

The ratio of the lengths of two corresponding diagonals of two similar polygons is equal to the ratio of the lengths of any two corresponding sides of the polygons.

Theorem 3

The perimeters of two similar polygons have the same ratio as the measures of any pair of corresponding sides of the polygons.

Theorem 4

Two polygons composed of the same number of triangles similar each to each, and similarly placed, are similar.

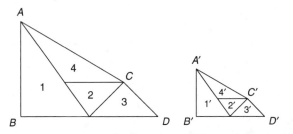

Problem Solving Examples:

Prove that any two regular polygons with the same number of sides are similar.

For any two polygons to be similar, their corresponding angles must be congruent and their corresponding sides proportional. It is necessary to show that these conditions always exist between regular polygons with the same number of sides.

Let us examine the corresponding angles first. For a regular polygon with n sides, the measure of each central angle is $\frac{360}{n}$ and each vertex angle is $\frac{(n-2)180}{n}$. Therefore, two regular polygons with the same number of sides will have corresponding central angles and vertex angles that are all of the same measure and, hence, are all congruent. This fulfills our first condition for similarity.

We must now determine whether or not the corresponding sides are proportional. It will suffice to show that the ratios of the lengths of every pair of corresponding sides are the same.

Since the polygons are regular, the sides of each one will be equal. Call the length of the sides of one polygon ℓ_1 and the length of the sides of the other polygon ℓ_2. Hence, the ratio of the lengths of corresponding sides will be ℓ_1/ℓ_2. This will be a constant for any pair of corresponding sides and, hence, the corresponding sides are proportional.

Thus, any two regular polygons with the same number of sides are similar.

Q The lengths of two corresponding sides of two similar polygons are 4 and 7. If the perimeter of the smaller polygon is 20, find the perimeter of the larger polygon.

A We know, by theorem, that the perimeter of two similar polygons have the same ratio as the measures of any pair of corresponding sides.

If we let s and p represent the side and perimeter of the smaller polygon and s' and p' the corresponding side and perimeter of the larger one, we can then write the proportion

$$p : p' = s : s'$$

By substituting the given, we can solve for p'.

$$20 : p' = 4 : 7$$
$$4p' = 140$$
$$p' = 35.$$

Therefore, the perimeter of the larger polygon is 35.

Properties of Similarity

Reflexive property

Polygon $ABCD \sim$ Polygon $ABCD$.

Symmetric property

If polygon $ABCD \sim$ polygon $EFGH$, then polygon $EFGH \sim$ polygon $ABCD$.

Transitive Property

If polygon *ABCD* ~ polygon *EFGH*, and polygon *EFGH* ~ polygon *WXYZ*, then polygon *ABCD* ~ polygon *WXYZ*.

Problem Solving Examples:

 Show that a regular pentagon and the pentagon determined by joining all the vertices of the regular pentagon are similar.

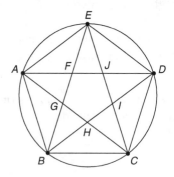

 We must show that pentagon *FGHIJ* ~ polygon *ABCDE*. Polygon *ABCDE* is a regular pentagon. Therefore, *FGHIK* must also be a regular pentagon. Since all regular pentagons are similar, it is sufficient to show that *FGHIJ* is a regular pentagon to complete the proof.

To show *FGHIJ* is a regular pentagon, we show that all angles are congruent and all sides are congruent.

The sides of pentagon *FGHIJ* are corresponding sides of △*AGF*, △*EFJ*, △*DJI*, △*CIH*, and △*BGH*. We show these triangles congruent by the SAS Postulate. The interior angles of the pentagon have opposite angles and, as such, angles congruent to them in △*AFE*, △*EJD*, △*DIC*, △*CHB*, and △*BGA*. We show these triangles congruent by ASA.

In our proof, we (1) show triangles formed by three consecutive vertices are congruent—that is △*AED*, △*EDC*, △*DCB*, △*CBA*, and △*BAE* by SAS.

(2) Therefore, $\measuredangle ABE \cong \measuredangle EAD \cong \measuredangle AEB \cong \measuredangle DEC \cong \ldots$; and

(3) $\triangle BGA \cong \triangle AFE \cong \triangle EJD \cong \triangle DIC \cong \triangle CHB$ by ASA.

By corresponding angles, the angles opposite the interior angles are congruent. Thus, all five interior angles are congruent.

To show the sides congruent, we (4) show $\overline{CI} \cong \overline{CH} \cong \overline{DJ} \cong \overline{DI} \cong$ $\overline{BG} \cong \overline{BH} \cong \overline{AG} \cong \overline{AF} \cong \overline{EF} \cong \overline{EJ}$ by corresponding parts.

(5) Show $\triangle ACD \cong \triangle EBC \cong \triangle DAB \cong \triangle CAE \cong \triangle BED$ by SSS; (6) Therefore $\measuredangle DAC \cong \measuredangle BEC \cong \measuredangle ADB \cong \measuredangle ECA \cong \measuredangle DBE$; and (7) $\triangle AGF \cong \triangle EFJ \cong \triangle DJI \cong \ldots$; and (8) by corresponding sides GF $\cong$ FJ $\cong$ JI $\cong$ $\overline{IH} \cong \overline{GH}$.

Given: regular pentagon *ABCDE*.
$\overline{AD} \cap \overline{EB} =$ point F; $\overline{AD} \cap \overline{EJ} =$ point J.
$\overline{AC} \cap \overline{BE} =$ point G; $\overline{AC} \cap \overline{BD} =$ point H.
$\overline{BD} \cap \overline{CD} =$ point I.

Prove: pentagon *FGHIJ* ~ *ABCDE*.

Statement	Reason
1. (see above)	1. Given.
2. $\overline{AE} \cong \overline{ED} \cong \overline{CD} \cong \overline{CB} \cong \overline{BA}$	2. The sides of a regular polygon are congruent.
3. $\measuredangle EAB \cong \measuredangle AED \cong \measuredangle EDC \cong$ $\measuredangle DCB \cong \measuredangle CBA$	3. The angles of a regular polygon are congruent.
4. $\triangle BAE \cong \triangle AED \cong \triangle EDC \cong$ $\triangle DCB \cong \triangle CBA$	4. The SAS Postulate.
5. $\measuredangle ABE \cong \measuredangle EAD \cong \measuredangle DEC \cong$ $\measuredangle CDB \cong \measuredangle BCA$ Also, $\measuredangle AEB \cong \measuredangle EDA \cong$ $\measuredangle DCE \cong \measuredangle CBD \cong \measuredangle BAC$	5. Corresponding parts of congruent triangles are congruent.
6. $\triangle BAE, \triangle AED, \triangle EDC,$ $\triangle DCB, \triangle CBA$ are isosceles	6. If two sides of a triangle are congruent, then the triangle is isosceles.
7. $\measuredangle ABE \cong \measuredangle EAD \cong \measuredangle DEC \cong$ $\measuredangle CDB \cong \measuredangle BCA \cong \measuredangle AEB \cong$ $\measuredangle EDA \cong \measuredangle DCE \cong \measuredangle CBD \cong$ $\measuredangle BAC$	7. The base angles of an isosceles triangle are congruent. Also, step 5.

8. $\triangle AFE \cong \triangle EJD \cong \triangle DIC \cong$ $\triangle CHB \cong \triangle BGA$

8. The ASA Postulate.

9. $\angle AFE \cong \angle EJD \cong \angle DIC \cong$ $\angle CHB \cong \angle BGA$

9. Corresponding angles of congruent triangles are congruent.

10. $\angle GFJ \cong \angle AFE$ $\angle FJI \cong \angle EJD$ $\angle JIH \cong \angle DIC$ $\angle GHI \cong \angle BHC$ $\angle FGH \cong \angle AGB$

10. Opposite angles are congruent.

11. $\angle GFJ \cong \angle FJI \cong \angle JIH \cong$ $\angle GHI \cong \angle FGH$

11. Transitivity Postulate. (Thus, the interior angles are congruent.)

12. $\triangle AFE$, $\triangle EJD$, $\triangle DIC$, $\triangle CHB$, and $\triangle BGH$ are isosceles triangles

12. If two angles of a triangle are congruent, then the triangle is isosceles. (see Step 7)

13. $\overline{AF} \cong \overline{EF} \cong \overline{EJ} \cong \overline{DJ} \cong \overline{DI} \cong$ $\overline{CI} \cong \overline{CH} \cong \overline{BH} \cong \overline{BG} \cong \overline{AG}$

13. In an isosceles triangle, the sides opposite the base angles are congruent.

14. $\overline{AD} \cong \overline{AC} \cong \overline{EB} \cong \overline{EC} \cong \overline{BD}$

14. Corresponding sides of congruent triangles are congruent. (see Step 4)

15. $\triangle CAD \cong \triangle BEC \cong \triangle ADB \cong$ $\triangle ECA \cong \triangle EBD$

15. The SSS Postulate.

16. $\angle CAD \cong \angle BEC \cong \angle ADB \cong$ $\angle ECA \cong \angle EBD$

16. The corresponding angles of congruent triangles are congruent.

17. $\triangle GAF \cong \triangle FEJ \cong \triangle JDI \cong$ $\triangle ICH \cong \triangle HBG$

17. The SAS Postulate.

18. $\overline{GF} \cong \overline{FJ} \cong \overline{JI} \cong \overline{IH} \cong \overline{HG}$

18. Corresponding parts of congruent triangles are congruent.

19. Pentagon *FGHIJ* is a regular pentagon

19. If the sides of a polygon are all congruent and the angles are all congruent, then the polygon is regular.

20. Pentagon *FGHIJ* ~ *ABCDE*

20. All regular polygons of the same number of sides are similar.

Quiz: Geometric Proportions and Similarity

1. The two triangles shown are similar. Find $m \angle 1$.

   (A) 48°

   (B) 53°

   (C) 74°

   (D) 127°

   (E) 180°

2. The two triangles shown are similar. Find a and b.

   (A) 5 and 10

   (B) 4 and 8

   (C) $4\frac{2}{3}$ and $7\frac{1}{3}$

   (D) 5 and 8

   (E) $5\frac{1}{3}$ and 8

3. The perimeter of $\triangle LXR$ is 45 and the perimeter of $\triangle ABC$ is 27. If $\overline{LX} = 15$, find the length of $\overline{AB}$.

   (A) 9

   (B) 15

   (C) 27

   (D) 45

   (E) 72

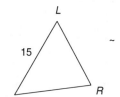

4. Find *b*.

    (A) 9

    (B) 15

    (C) 20

    (D) 45

    (E) 60

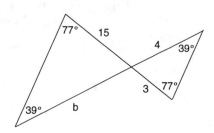

5. Find the area of Δ*MNO*.

    (A) 22

    (B) 49

    (C) 56

    (D) 84

    (E) 112

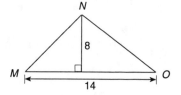

6. In the figure, sin∢*BCD* = $\frac{1}{2}$, $\overline{BC}$ = 8, and ∢*BDC* = 90°. The ratio of the area of Δ*ABD* to Δ*DBC* is

    (A) $\frac{2}{3}$.

    (B) $\frac{4}{5}$.

    (C) $\frac{1}{4}$.

    (D) $\frac{1}{2}$.

    (E) $\frac{1}{3}$.

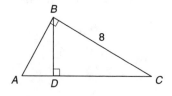

7. In the figure, if $AC \parallel DF$ and if $DE = a$ and $EF = b$, then $\dfrac{AB}{AC} =$

    (A) $\dfrac{a}{b}$.

    (B) $\dfrac{a}{a+b}$.

    (C) $\dfrac{a}{b} - 1$.

    (D) $\dfrac{b}{a}$.

    (E) $\dfrac{b}{a} - 1$.

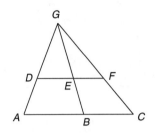

8. In the figure below, if $CD \parallel AB$ and $\dfrac{CD}{AB} = \dfrac{1}{2}$, then what is the ratio of the area of $\triangle CED$ to the area of $\triangle AEB$?

    (A) $\dfrac{1}{8}$

    (B) $\dfrac{1}{4}$

    (C) $\dfrac{1}{2}$

    (D) 2

    (E) 4

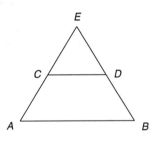

9. In the figure below, if $AC = BC = CD$ and $\sphericalangle CAD = \sphericalangle CBD = 15°$ then $\theta =$

    (A) $15°$.

    (B) $30°$.

    (C) $45°$.

    (D) $60°$.

    (E) $75°$.

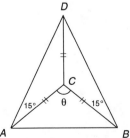

10. In the figure below, if $BC \parallel DE$ and if $AE = a$ and $AC = b$, then $\dfrac{BC}{DE} =$

(A) $\dfrac{b}{a}$.

(B) $1 - \dfrac{b}{a}$.

(C) $1 + \dfrac{b}{a}$.

(D) $a - b$.

(E) None of the above

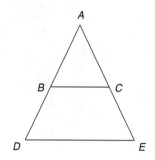

ANSWER KEY

1.	(B)	6.	(E)
2.	(E)	7.	(B)
3.	(A)	8.	(B)
4.	(C)	9.	(D)
5.	(C)	10.	(A)

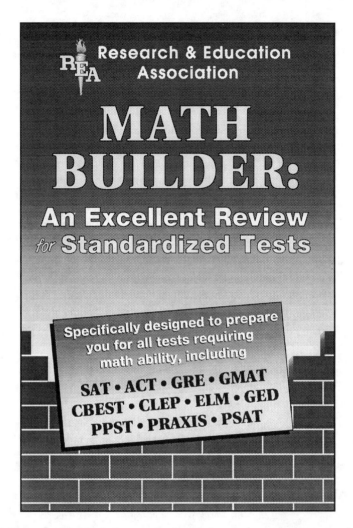